NOTHING IS FIXED EVERYTHING CHANGES

Simple Mathematical and Physical Models for Different Types of Changes

by

M. Kemal Atesmen

2018

Archway Publishing books may be ordered through booksellers or by contacting:

Archway Publishing
1663 Liberty Drive
Bloomington, IN 47403
www.archwaypublishing.com
1 (888) 242-5904

ISBN: 978-1-4808-7080-2 (sc)
ISBN: 978-1-4808-7079-6 (e)

Library of Congress Control Number: 2018913009

Print information available on the last page.

Archway Publishing rev. date: 11/20/2018

Table of Contents

Preface

Everything changes in our universe and in our world. Our universe is expanding continuously. Our genomes change constantly to respond to new living conditions. Our power of remembering things declines slowly as we age. A raindrop evaporates, reduces in size and disappears. Pressure increases as we dive to the depths of an ocean. A radioactive material decays by time and loses its mass. A sprinter accelerates to increase his or her speed to finish the race. Based on long term and reliable observations, we can approximate changes in the behavior of a dependent variable such as mass history of a radioactive material, pressure in the depths of an ocean, speed of an athlete, etc., with respect to an independent variable such as time, depth, etc. with mathematical modeling.

Mathematical modeling of change can be a linear one that only requires knowledge of algebra. More complicated change models require knowledge of calculus, statistics, probability, first order or higher order linear or non-linear differential equations. I will use simple models to approximate behavior of twenty-four different change phenomena. Most mathematical models used in this book are first order difference (change) equations. In several models, I used statistics and probability to approximate a change.

In the present digital world, one can approximate a change by using reliable historical data and proven and complicated mathematical models to forecast changes in weather, in stock prices, in future inflation values, etc. Mathematical models of change presented in this book are simple and straightforward. Most of the time a change in the dependent variable depends only on one earlier value of itself. So I would like to recommend this book as an introductory mathematical modeling for change.

Acknowledgements

Over 40 years of engineering, engineering management and project management in the global arena covering automotive, computer, data communication, and offshore oil industries were accomplished by exceptional support from my wife, Zeynep, and my family members. Some years I was away from home more than six months out of a year trying to tackle challenging engineering tasks.

I would like to dedicate this book to all engineering project teams' members with whom I had the pleasure of working together over the years, with enthusiasm, with imagination and with determination. Over these years my engineering project teams' members kept coming back to work along with me without any reservations.

M. Kemal Atesmen

Santa Barbara, California

About the Author

M. Kemal Atesmen completed his high school studies at Robert Academy in Istanbul Turkey in 1961. He received his B. Sc. degree from Case Western Reserve University, his M. Sc. degree from Stanford University, and his Ph. D. degree from Colorado State University, all in mechanical engineering. He is a life member of ASME. He initially pursued an academic and an industrial career in parallel and became an associate professor in mechanical engineering before dedicating his professional life to international engineering management and engineering project management for thirty-three years. He helped many young engineers in the international arena to bridge the gap between college and professional life in automotive, computer component, data communication, and offshore oil industries.

He published eight books, sixteen technical papers, and has four patents. His books are "Global Engineering Project Management", CRC Press, 2008, "Everyday Heat Transfer Problems – Sensitivities to Governing Variables", ASME Press, 2009, "Understanding the World Around through Simple Mathematics", Infinity Publishing, 2011, "A Journey Through Life, Wilson Printing, 2013, "Project Management Case Studies and Lessons Learned", CRC Press, 2015, "Process Control Techniques for High Volume Production", CRC Press, 2016, "Engineering Management in a Global Environment: Guidelines and Procedures", CRC Press, 2017, and "Case Studies in Fluid Mechanics with Sensitivities to Governing Variables", Wiley and ASME Press, 2018.

Introduction

Everything changes continuously in our universe, in our world and in ourselves. There are many mathematical tools to describe these changes from observations, from collection of historical data, from experimental data and from our vision. Mathematical tools such as algebra, calculus, differential equations, probability and statistics help us to model a changing dependent variable with respect to governing independent variables. Depth of knowledge about a change, a correct mathematical model with sound assumptions and reliable data about that change create a trustworthy understanding of that change in us.

In this book, seventeen different physical or everyday life phenomena, namely seventeen chapters, will concentrate on zeroth order and first order changes, namely changes that can be modeled using zeroth order and first order finite difference and differential equations. Five of the problems focus on optimization of a dependent variable with respect to an independent variable using simple algebraic equations. One problem investigates changes in a process or in a measurement system in high volume production using industrial statistics. In another problem, I investigate changes to dependent variables using probability analysis.

There are several reference books that the reader can refer to for further understanding of several branches of physics, mathematics, fluid mechanics and heat transfer that model a change. One such book is in differential equations, namely "Elementary Differential Equations and Boundary Value Problems" by W. E. Boyce and R. C. DiPrima, Wiley, 1997, Reference [2]. Another excellent reference book in physics is "Physics" by D. Halliday, R. Resnick and K. S. Krane, Wiley, 1992, Reference [3]. Two excellent books on industrial statistics and probability are "Statistics for Management" by D. Levin, Prentice-Hall, 1987, Reference [5] and "Process Control Techniques for High-Volume Production" by M. K. Atesmen, CRC Press, 2017, Reference [1]. My recommended reference book on changes in fluid flows is "Mechanic of Fluids" by I. H. Shames, McGraw-Hill, 1962, Reference [6]. My recommended book on changes in heat transfer is "Principles of Heat Transfer" by F. Kreith, International Textbook Company, 1965, Reference [4].

In Chapter 1, I investigate how the amount in a college fund changes until a student starts college for different interest rates and for different compounding modes.

Chapter 2 models the change in temperature of coffee in a cup as time changes. Temperature of the coffee in a cup is proportional to the heat flow from the coffee and the cup to their environment.

Everything we learn does not retain in our brains. Chapter 3 deals with decay of knowledge and memory in our brains in time. There is a continuous change in retained memory with respect to changing time.

Chapter 4 models changes in the motion of a car, a person, or a particle in a straight line and in one-dimension.

Chapter 5 analyzes a jumper on a trampoline who can experience changes in his or her vertical position, vertical speed and energy input from trampoline springs to the jumping mat while jumping up and down.

Vehicle decelerations depend heavily on tire surface to road surface traction characteristics. There are several different and sophisticated anti-skid braking systems that are used in vehicles today that prevent wheels to lock up during braking on any surface and improve vehicle control during braking. In Chapter 6, I investigate the stopping distance for a vehicle which has constant deceleration.

Chapter 7 looks into changes in jumping speeds and hang times for a jumper on Earth's surface and compares them to a jumper on Jupiter and on our Moon.

Chapter 8 models changes in the charge of a fully charged capacitor which is connected to a resistor in a RC circuit in which there is only a resistor and a fully charged capacitor.

In Chapter 9, the spread of a contagious disease in an isolated town is investigated. Outside effects due to travel and mixing of town's people with other region's people are not allowed in this analysis.

A raindrop's evaporation during its descend to the ground can be a very complicated physical phenomenon. Its shape, its temperature, surrounding's temperature, pressure, relative humidity and contaminations, wind conditions, solar radiation, and its impact with other raindrops can affect a raindrop's evaporation phenomenon. In Chapter 10, a simple model analyzes a raindrop's evaporation and its change in size with time.

Carbon monoxide, CO, release from a kerosene heater can be very dangerous for people and animals in an unventilated room. Chapter 11 analyzes the change in mass of carbon monoxide in an unventilated room with changing time.

In Chapter 12, changes in population of people and animals with respect to changing time can be predicted by product of a constant population change rate and the population present at the initial time. This type of analysis excludes sudden occurrences in population change due to natural disasters, epidemics, migrations, wars, etc.

As we go higher in altitude in our Earth's atmosphere, the density of air decreases so does the air pressure forces acting on our bodies. Changes in pressure as a result of changes in altitude is modeled in Chapter 13 by balancing out the pressure and gravitational forces acting on a body.

When we dive into the sea, the pressure around us increases very fast due to high density of sea water which is about 1,030 kg/m^3. Most divers should not go deeper than forty meters without special equipment where the sea water pressure reaches five atmospheres and then nitrogen narcosis sets in causing loss of sense and movement. Changes in pressure in water as a result of changes in depth can be modeled by balancing out the pressure and gravitational forces acting on a body as shown in Chapter 14.

Changes in mass of a radioactive material with respect to changing time is a first order chemical reaction and can be predicted by the product of a constant decay rate and the mass of the radioactive material present at initial time as analyzed in Chapter 15.

Chapter 16 models water flowing out of a small drain hole centered at the bottom of a large cylindrical tank. The analysis in this chapter assumes a frictionless and a non-rotating flow of an incompressible fluid, i.e. water, throughout the tank and the drain hole. Then the governing fluid mechanics equations simplify to the Bernoulli's principle along a fluid's streamline.

Chapter 17 models a bacteria which multiplies rapidly under right conditions and nutrients. The number of bacteria increases exponentially which can be modeled very similar to a continuously compounding interest.

Chapters 18 through 21 deal with optimization of a dependent variable by changing an independent variable. Chapter 18 optimizes the sales price of a new book. Sales volume of a new book can be very price sensitive. So in this chapter, a publishing company is going to release a new book and they want to price the book in such a way that they can maximize their revenues from its sales.

In most optimization problems, analysis of the first and second derivatives of a dependent variable with respect to an independent variable has to be done to find the minimum or the maximum value of the dependent variable. In order to illustrate such a problem, Chapter 19 considers a ladder that has to lean on a wall in order to reach a structure on the other side of the wall and tries to determine the shortest ladder length that is required to reach the structure.

In Chapter 20, a combination of long term heating costs and insulation costs for a house are analyzed in order to determine the optimum insulation thickness that will minimize long term heating plus insulation costs.

In Chapter 21, expansion of a vineyard is optimized. The quality of a wine starts in the vineyard. Vineyards are intentionally farmed to produce low yields so that they can get the best tasting grapes. In this case, the vineyard owner wants to add additional acreage to his vineyard in order to increase his good grape production and therefore his wine production in order to maximize his profits.

Chapter 22 analyzes a change in a process statistically. Changes in a variable that is critical to a process can be identified by collecting a set of controlled data and then analyzing it statistically. In this chapter, I analyze an egg farm where eggs under a certain size are categorized as rejects and cannot be shipped to customers. The egg farmer records meticulously, on a daily basis, number of rejected small eggs and his daily total egg production. He wants to improve his production and his shippable egg yields by performing a genetic manipulation to his egg laying hens.

Chapter 23 treats changes in a high volume production control of chip wafers for plating layer thickness. In high volume production, it is not feasible and cost effective to measure each part to see if it complies with a critical specification. First we have to have a capable measurement system for the critical specification in question in order to be able to measure samples in a production line. Control charting a critical specification by sampling can show you immediately, if your process is changing or it is out-of-control. Similar control charting applies to your measurement systems. You can make immediate decisions to shut a high volume production line down in order to find and correct the problem.

In this chapter, variable control charts such as $\bar{X}$, i.e. average of production samples, and R , range of production samples, are used to provide excellent information about the overall average and spread of a product with respect to a critical specification.

In Chapter 24, an event is characterized by the probability of its happening. Changes in the probability of an event can provide crucial information for decision making. Two events that are not related, namely mutually exclusive, and two events that are related, namely mutually non-exclusive, are investigated.

I would like to dedicate this book to my excellent teachers and mentors in mathematics, in physics, in fluid mechanics and in heat transfer at Robert Academy, at several universities and organizations. Some of the names at the top of a long list are Mr. Jacobson, Prof. I. Flugge-Lotz, Prof. W. C. Reynolds, Prof. W. M. Kays, Prof. A. L. London, Prof. R. D. Haberstroh, Prof. L. V. Baldwin and Prof. T. N Veziroglu.

Chapter 1

Saving for College Education

My dad started to put money into a college saving fund from the day I was born. He invested a fixed amount of money in the fund every year until I was nineteen years old. Let us see how the amount in my college fund changes until I start college for different interest rates and for different compounding modes.

The amount of money that I have in my college fund, M, will change over time, namely a change of (M_j-M_{j-1}) , during a change in time, (t_j-t_{j-1}). This change in the amount of money depends upon earnings from interest, i, during that change in time, namely (i x M_{j-1}) , plus the deposit, D, that my parents put into my college fund at the beginning of a time period. The change in the amount of money in my college fund with respect to changing time can be defined by a first order difference relationship shown in Equation 1-1.

$$\frac{M_j - M_{j-1}}{t_j - t_{j-1}} = i \times M_{j-1} + D \qquad\qquad 1\text{-}1$$

where *j-1* represents the previous time period (i.e. in years, months, days, etc.), and *j* represents the present time period.

The change in the amount of money in my college fund from the previous year to this year can be represented by the following simple linear Equation 1-2, if the time period $(t_j - t_{j-1})$ is assumed to be one year. It is also assumed that *i* is the interest rate compounded annually and D is the deposit to the fund at the beginning of each year.

$$M_j = M_{j-1} + i \times M_{j-1} + D \qquad\qquad \text{1-2}$$

If the interest is compounded semi-annually, namely $(t_j - t_{j-1})$ is a six month period. We have to use modified Equation 1-2 for every six month period as shown in Equations 1-3a and 1-3b. Assuming *i* is the annual interest rate compounded semi-annually and D is the deposit to the fund at the beginning of each year, we have to use the following two equations in their respective periods.

At the beginning of the year we use Equation 1-3a.

$$M_j = M_{j-1} + \left(\frac{i}{2}\right) \times M_{j-1} + D \qquad\qquad \text{1-3a}$$

At the end of six months and at the end of the year we use Equation 1-3b without the deposit.

$$M_j = M_{j-1} + \left(\frac{i}{2}\right) \times M_{j-1} \qquad\qquad \text{1-3b}$$

If the interest is compounded monthly, we have to use modified Equation 1-2 for every month. The variable i is the annual interest rate compounded monthly and D is the deposit to the fund at the beginning of each year. Modified Equation 1-2 for monthly compounded interest are given in Equations 1-4a and 1-4b. $(t_j - t_{j-1})$ is a one month period.

At the beginning of the year we use Equation 1-4a.

$$M_j = M_{j-1} + \left(\frac{i}{12}\right) \times M_{j-1} + D \qquad\qquad \text{1-4a}$$

At the end of every month we use Equation 1-4b without the deposit.

$$M_j = M_{j-1} + \left(\frac{i}{12}\right) \times M_{j-1} \qquad \text{1-4b}$$

If the interest is compounded daily, we have to use modified Equation 1-2 for every day as shown in Equations 1-5a and 1-5b. $(t_j - t_{j-1})$ is a one day period, i is the annual interest rate compounded daily, and D is the deposit to the fund at the beginning of each year.

At the beginning of the year, we use Equation 1-5a.

$$M_j = M_{j-1} + \left(\frac{i}{365}\right) \times M_{j-1} + D \qquad \text{1-5a}$$

For other days of the year, we use Equation 1-5b without the deposit.

$$M_j = M_{j-1} + \left(\frac{i}{365}\right) \times M_{j-1} \qquad \text{1-5b}$$

As the time period $(t_j - t_{j-1})$ get smaller and smaller and approaches zero, first order difference Equation 1-1, $\frac{M_j - M_{j-1}}{t_j - t_{j-1}} = i \times M_{j-1} + D$, becomes a first order differential equation as shown in Equation 1-6.

$$\frac{dM}{dt} = i \times M + D \qquad\qquad 1\text{-}6$$

If my parents deposit an amount D_o to my college fund at $t = 0$, the solution to Equation 1-6 is given in Equation 1-7 for continuous compounding with an interest rate of i and a constant deposit rate of D.

$$M = \left[D_o + \left(\tfrac{D}{i}\right)\right] \times e^{i \times t} - \left(\tfrac{D}{i}\right) \qquad\qquad 1\text{-}7$$

Using Equation 1-7, my college fund amount after 18 years is shown in Figure 1-1 for continuous compounding of different interest rates with an initial deposit of $2,000 and annual deposits of $2,000. The amount accumulated from interest in 18 years takes over the amount deposited in 18 years around 6.9% interest rate.

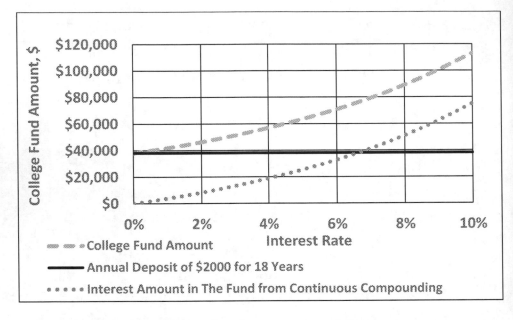

Figure 1-1: College fund amounts with continuous compounding of different interest rates for 18 years with a deposit of $2,000 per year

Using Equations1-2a through 1-5b, we can calculate the amount of college fund that accumulates for different compounding modes and for continuous compounding for 18 years. Results that are presented in Table1-1 are for a 10% interest rate, for an initial deposit of $2,000 and an annual deposit of $2,000.

Interest Compounding Mode	College Fund Amount in 18 Years
1 Year	$102,318
6 Months	$105,082
1 Day	$107,598
Continuous Compounding	$113,092

Table 1-1 College fund accumulation for 18 years for different interest compounding modes

Chapter 2

Cooling of Coffee in a Cup

When we pour hot coffee into a cup, the coffee in the cup cools down and approaches the temperature of its environment as time goes by. The change in the temperature of the coffee as time changes is proportional to the heat flow from the coffee and the cup to their environment. The change in the temperature of the coffee in the cup with respect to changing time can be defined by a first order difference relationship shown in Equation 2-1.

$$\frac{T_j - T_{j-1}}{t_j - t_{j-1}} = -K \times \left(T_{j-1} - T_{env} \right) \qquad \text{2-1}$$

where *j-1* represents the previous time and *j* represents the present time. T represents the average temperature of the coffee in the cup in centigrade. The variable *t* is the time in seconds. T_{env} is the average temperature of the environment around the cup in centigrade and it is assumed to be a constant.

K is the proportionality constant in 1/second. Definition of K is given in Equation 2-2. The minus sign in Equation 1-1 shows that the coffee temperature in the cup decreases as time goes by when $(T_{j-1}-T_{env}) > 0$.

$$K = \frac{h \times A}{c_p \times \rho \times V} \qquad\qquad 2\text{-}2$$

where h is the heat transfer coefficient for coffee's cooling process in $W/(m^2\text{-}°K)$. The heat transfer coefficient can be obtained experimentally. It is assumed to be a constant in the present calculations, and it can encompass convective, radiative and conductive modes of heat transfer. The surface area of the cup is A in m^2 from which heat is transferred to the environment. The variable c_p is the specific heat of the coffee in the cup at constant pressure in $J/(kg\text{-}°K)$. The variable ρ is the density of the coffee in the cup in kg/m^3. V is the volume of the coffee in the cup in m^3.

In order to determine the proportionality constant K, the following parameters are used:

$$h = 40 \frac{W}{m^2 - {}^0K} \; for \; an \; uninsulated \; coffee \; cup$$

$$A = 0.034 \; m^2$$

$$c_p = 4200 \; \frac{J}{kg - {}^0K}$$

$$\rho = 970 \; \frac{kg}{m^3}$$

$$V = 0.00035 \; m^3$$

$$T_o = 90 \; {}^0C \; (initial \; coffee \; temperature \; in \; the \; cup)$$

$$T_{env} = 20 \; {}^0C$$

The proportionality constant becomes as given in Equation 2-3.

$$K = 0.00095 \; \frac{1}{s} \hspace{4cm} \text{2-3}$$

(1/K) is also called a time constant. In this case, using Equation 2-3, the time constant is 1,053 seconds or 17.5 minutes.

In 17.5 minutes, the liquid temperature in the cup will drop down to 36.8% of the initial temperature difference, namely 36.8% of $(T_o - T_{env}) = 90 - 20 = 70\ ^oC$. In 17.5 minutes, the liquid temperature will be $(20 + 0.368 \times 70) = 46\ ^oC$.

If we consider an insulated coffee cup, the heat transfer coefficient reduces to the following value:

$$h = 17 \frac{W}{m^2 - {}^oK}\ for\ an\ insulated\ coffee\ cup$$

Then for an insulated coffee cup, the proportionality constant K becomes 0.00041 1/s and the time constant increases to 41 minutes; that is, the coffee cools down to 46 ^oC in 41 minutes in an insulated cup. Using Equation 2-1, a delta t of 10 seconds and parameters given above for an insulated and an uninsulated coffee cup, coffee temperature versus time plots are shown in Figure 2-1.

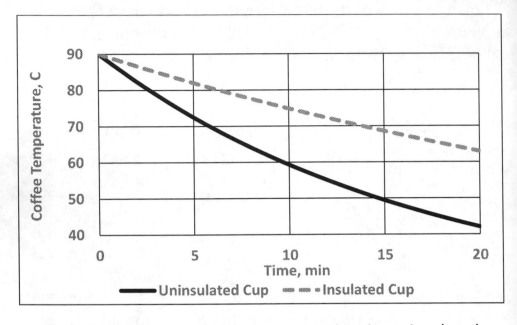

Figure 2-1: Coffee temperature versus time in an insulated and in an uninsulated cup

As the time change $(t_j - t_{j-1})$ get smaller and smaller and approaches zero, first order difference Equation 2-1, $\frac{T_j - T_{j-1}}{t_j - t_{j-1}} = -K \times (T_{j-1} - T_{env})$, becomes a first order differential equation as shown in Equation 2-4.

$$\frac{dT}{dt} = -K \times (T - T_{env}) \qquad\qquad 2\text{-}4$$

Equation 2-4 can be integrated, and then by applying the initial coffee temperature condition of T_o at $t = 0$, the following Equation 2-5 is obtained as the exact solution.

$$T = T_{env} + (T_o - T_{env}) \times e^{-K \times t} \qquad \text{2-5}$$

Change in coffee temperature with respect to changing time can accurately be obtained using the first order difference relationship given in Equation 2-1 as long as small enough delta t's are utilized in the calculations. Figure 2-2 compares coffee temperatures in an uninsulated cup versus time from the exact solution, namely using Equation 2-5, to the first order difference relationship given in Equation 2-1. As shown in Figure2-2, large time changes, i.e. 5 minutes, do deviate from the exact solution. As delta t get below 1 minute, first order difference relationship follows the exact solution accurately.

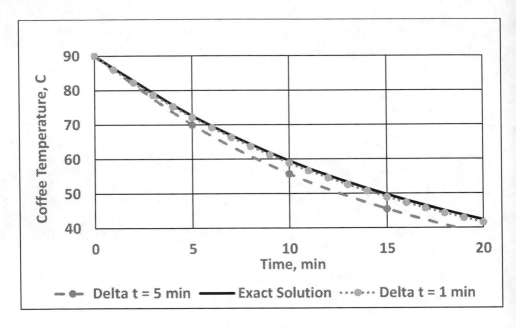

Figure 2-2: Change in coffee temperature
versus time for different delta *t*'s
compared to exact solution

Chapter 3

Memory Retention

If we want to be successful in life, we have to learn new skills, new information, new languages, etc. The downside of learning is that everything we do not retain everything we learn. Knowledge and memory in our brains decay in time. There is a continuous change in retained memory with respect to changing time. The first scientist to work on change in memory retention was the German psychologist Hermann Ebbinghaus in the 19th century. He performed lots of memory experiments and showed that a change in memory retention with respect to changing time is proportional to the memory retained in our brains. A change in memory retention with respect to changing time is given in Equation 3-1 as a first order difference relationship.

$$\frac{M_j - M_{j-1}}{t_j - t_{j-1}} = -r \times M_{j-1} \qquad \text{3-1}$$

The minus sign in Equation 3-1 comes from the fact that our memory retention always declines. M is the amount of information retained in our brain at a certain time t.

The variable j represents the present time. Previous time is represented by $j-1$. The variable r is the memory retention rate. $(1/r)$ is a time constant τ, which represents the strength of your memory retention. Large time constant means that you retain more of your memory as time goes by. Percent memory retention as a function of time is shown in Figure 3-1 using Equation 3-1 and using three different τ values, namely 7 days, 30 days and 90 days. As τ gets larger, we retain more of our memory. For example, we retain 50% of the information that we put into our brains after 4.5 days, if we have a memory retention strength of 7 days. We retain 50% of the information that we put into our brains after 20 days, if we have a memory retention strength of 30 days. We retain 50% of the information that we put into our brains after 62 days, if we have a memory retention strength of 90 days

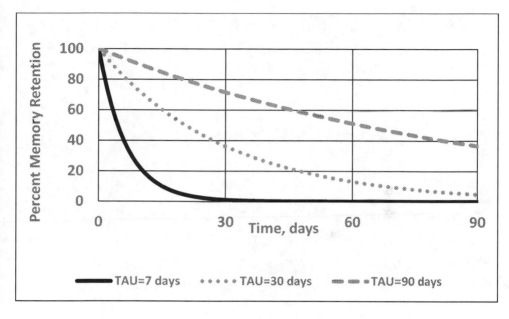

Figure 3-1: Percent memory retention versus time

As the time change $(t_j - t_{j-1})$ get smaller and smaller and approaches zero, first order difference Equation 3-1 becomes a first order differential equation as shown in Equation 3-2.

$$\frac{dM}{dt} = -r \times M \qquad\qquad\text{3-2}$$

If we call the initial information retained in our memory M_o, the exact solution to Equation 3-2 is shown in Equation 3-3.

$$M = M_o \times e^{-r \times t} \ or \ M = M_o \times e^{-\frac{t}{\tau}} \qquad 3\text{-}3$$

If a small enough time change $(t_j - t_{j-1})$ is used in calculations for memory retention, i.e. less than a day, results obtained by using Equation 3-1 are very much in line with results obtained from the exact solution of Equation 3-3.

If an information is put into our memory several times in different intervals, this repeated information stays in our memory longer, namely its τ increases. For example, if we are studying for a biology examination which requires a strong memory retention, we might have to repeat our studies for this examination a couple of times at different intervals. Such a memory retention case is analyzed using Equation 3-1 and is presented in Figure 3-2. Initial study has a τ of 2 days and the examination will take place after 10 days from the initial study. Second study is performed 2 days after the initial study and it has a τ of 7 days. The third study is performed 4 days after the second study and it has a τ of 21 days.

During the examination day, the student has a memory retention of 80% which shows the strengthening of memory for his or her biology examination material by repeated studies.

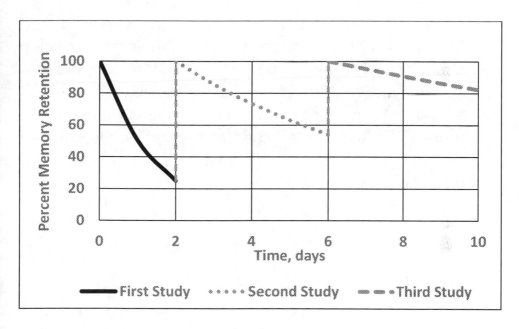

Figure 3-2: Memory retention curves for repeated studies

Chapter 4

Motion in a Straight Line

Let us investigate changes in motion of a car, a person, a particle, etc. in a straight line and in one-dimension. Consider a car accelerating from a stop sign to enter a freeway in a straight line motion. The driver steps on the accelerator pedal and his or her acceleration, A , increases so does the car's speed, S , as time, t , goes by. Average acceleration is defined as the change in speed with a change in time as shown in Equation 4-1.

$$\bar{A} = \frac{S_{j+1} - S_J}{t_{j+1} - t_j} \qquad\qquad 4\text{-}1$$

The car is at a speed of S_j at time t_j and at a speed of S_{j+1} at time t_{j+1} . As the time interval $(t_{j+1} - t_j)$ gets smaller and approaches zero, we can obtain the instantaneous acceleration as the derivative of speed with respect to time as given in Equation 4-2.

$$A = \frac{dS}{dt}$$ 4-2

A typical car's acceleration from a stop sign to enter a freeway in a straight line motion is shown in Figure 4-1.

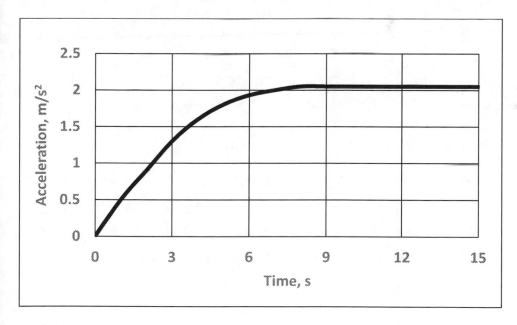

Figure 4-1: Car's acceleration versus time

Near the Earth's surface, the gravitational acceleration is 9.81 m/s² which is mostly referred to as 1 G of acceleration. So the maximum acceleration of the car represented in Figure 4-1 is (2/9.81) G's or 0.2 G's.

Using the acceleration-time curve in Figure 4-1 along with Equation 4-1, we can calculate the increasing speed of the car. Equation 4-1 can be rewritten to determine the speed of the car at time t_j as shown in Equation 4-3.

$$S_j = S_{j-1} + \left(\frac{A_j + A_{j-1}}{2}\right) \times \left(t_j - t_{j-1}\right) \qquad\qquad \text{4-3}$$

Actually we are integrating approximately the area under the acceleration- time curve in Figure 4-1 in order to obtain the speed of the car. The average acceleration of the car during the time interval $\left(t_{j+1} - t_j\right)$ is approximated by $\bar{A} = \left(\frac{A_j + A_{j-1}}{2}\right)$. As the time interval $\left(t_{j+1} - t_j\right)$ gets smaller and approaches zero, the calculated speed of the car will be more accurate. Car's speed versus time using 1 second time intervals is shown in Figure 4-2.

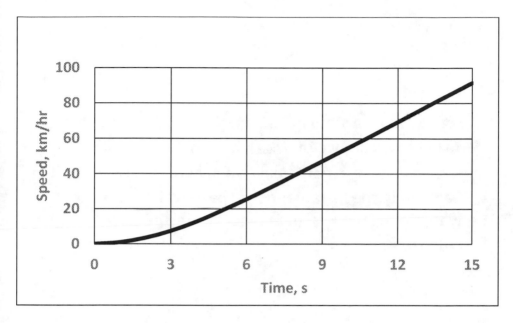

Figure 4-2: Car's speed versus time

The car that is accelerating on a straight line to enter the freeway, as in Figure 4-1, reaches a speed of 91.5 km/hr in 15 seconds.

From the speed versus time curve given in Figure 4-2, we can now determine the distance, D , versus, t , time curve. Average speed is defined as the change in a straight line distance with a change in time as shown in Equation 4-4.

$$\bar{S} = \frac{D_{j+1} - D_j}{t_{j+1} - t_j}$$ 4-4

The car traveled a straight line distance of D_j at time t_j and a straight line distance of D_{j+1} at time t_{j+1} . As the time interval $(t_{j+1} - t_j)$ gets smaller and approaches zero, then we can obtain the instantaneous speed of the car as the derivative of distance with respect to time as given in Equation 4-5.

$$S = \frac{dD}{dt}$$ 4-5

As the time interval $(t_{j+1} - t_j)$ gets smaller and approaches zero, using Equation 4-5, we can rewrite Equation 4-2 for the instantaneous acceleration of the car as the second derivative of distance with respect to time as given in Equation 4-6.

$$A = \frac{dS}{dt} = \frac{d}{dt}\left(\frac{dD}{dt}\right) = \frac{d^2D}{dt^2}$$ 4-6

Using the speed-time curve given in Figure 4-2 along with Equation 4-4, we can calculate the straight line distance traveled by the car. Equation 4-4 can be rewritten to determine the straight line distance traveled by the car at time t_j as shown in Equation 4-7.

$$D_j = D_{j-1} + \left(\frac{S_j + S_{j-1}}{2}\right) \times \left(t_j - t_{j-1}\right) \qquad\qquad 4\text{-}7$$

Actually we are at this step integrating approximately the area under the speed-time curve in Figure 4-2 in order to obtain the straight line distance traveled by the car. The average speed of the car during the time interval $\left(t_{j+1} - t_j\right)$ is approximated by $\bar{S} = \left(\frac{S_j + S_{j-1}}{2}\right)$. As the time interval $\left(t_{j+1} - t_j\right)$ gets smaller and approaches zero, the calculated distance of the car will be more accurate. Straight line distance traveled by the car versus time using 1 second time intervals is shown in Figure 4-3.

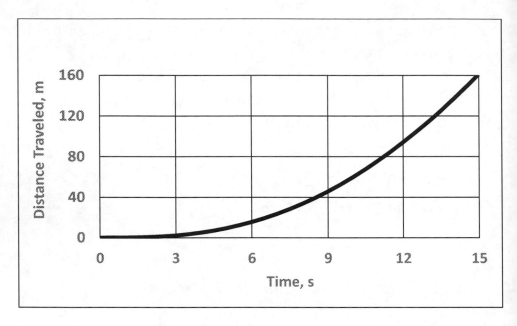

Figure 4-3: Distance traveled by the car on a straight line from a stop sign to 91.5 km/hr on the freeway

The car travels on a straight line a distance of 160 meters while reaching 91.5 km/hr speed with the acceleration-time profile given in Figure 4-1.

When a car starts to accelerate, we feel a jerk motion. Each time the car's acceleration changes, we experience the jerk motion. Jerk motion is defined as the time rate of change of acceleration in the direction of acceleration.

Average jerk, $\overline{JK}$, is defined as the change in acceleration with a change in time as shown in Equation 4-8. Jerk has units of m/s^{3}or sometimes G/s.

$$\overline{JK} = \frac{A_{j+1}-A_J}{t_{j+1}-t_j}$$ 4-8

In the present example, the car is at an acceleration of A_j at time t_j and at an acceleration of A_{j+1} at time t_{j+1} . As the time interval $(t_{j+1} - t_j)$ gets smaller and approaches zero, we can obtain the instantaneous jerk as the derivative of acceleration with respect to time as given in Equation 4-9.

$$JK = \frac{dA}{dt} = \frac{d^2S}{dt^2} = \frac{d^3D}{dt^3}$$ 4-9

Jerk motion experienced in the direction of the car traveling in a straight line versus time using 1 second time intervals is obtained from the acceleration-time curve given in Figure 4-1 and the results are shown in Figure 4-3.

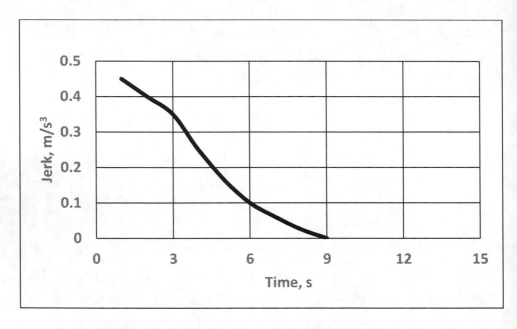

Figure 4-4: Car's average jerk versus time

The jerk motion is high at the beginning of accelerating car. As the change in acceleration decreases so does the jerk motion. The jerk motion is zero when the car reaches a constant acceleration mode.

Chapter 5

Jumping on a Trampoline

A jumper on a trampoline can experience changes in his or her vertical position, vertical speed and energy input from trampoline springs to the jumping mat while jumping up and down. In the present analysis, let us consider the changing vertical position and vertical speed of a jumper while neglecting the air drag resistance on the jumper. Let us assume that the energy input from trampoline springs to the jumping mat and then to the jumper is at a maximum and it is a constant when the center of the jumping mat is at its lowest point. When the springs are fully extended and the center of the jumping mat is at its lowest point with the jumper on this center point, let us call this lowest center point on the mat the zero datum or the $Z = 0$ point for the positive vertical direction.

When the jumper is airborne, the only force acting on the jumper is the downward gravitational force. When the jumper is airborne, his or her vertical speed will change with respect to changing time as given in Equation 5-1.

$$\frac{S_i - S_{i-1}}{t_i - t_{i-1}} = -g$$

5-1

where $g = 9.81 \ m/s^2$ is the gravitational acceleration on the surface of the Earth and it is a constant with respect to changes in time. As the time interval $(t_i - t_{i-1})$ gets smaller and approaches zero, we can re-write Equation 5-1 for the instantaneous speed of the jumper as the first derivative of vertical speed with respect to time as given in Equation 5-2.

$$\frac{dS}{dt} = -g$$

5-2

The change in the jumper's vertical position with respect to changing time can be obtained from the jumper's vertical speed-time profile with the following Equation 5-3.

$$\frac{Z_i - Z_{i-1}}{t_i - t_{i-1}} = \bar{S} = \frac{S_i + S_{i-1}}{2}$$

5-3

where $\bar{S}$ is the jumper's average vertical speed during the time interval $(t_i - t_{i-1})$.

As the time interval $(t_i - t_{i-1})$ gets smaller and approaches zero, we can rewrite Equation 5-3 for the instantaneous vertical position of the jumper as the first derivative of vertical position with respect to time as shown in Equation 5-4.

$$\frac{dZ}{dt} = S \qquad\qquad 5\text{-}4$$

Initially at $Z = 0$, it is assumed that the spring energy inputted the by the springs to the trampoline mat is totally transferred to the jumper's initial kinetic energy. This transfer of energy is shown in Equation 5-5.

$$0.5 \times k \times X^2 = 0.5 \times M \times S_o^2 \qquad\qquad 5\text{-}5$$

where k is the spring constant in N/m, X is the spring displacement in meters, M is the jumper's mass in kilograms and S_o is the initial vertical speed of the jumper.

In the present analysis, let us use the following parameters to determine the changing vertical position and vertical speed of a jumper.

$k = 20,000 \ N/m$

$X = 0.2 \ m$

$M = 50 \ kg$

Using Equation 5-5, it can be determined that the jumper has an initial vertical speed of 4 m/s or $S_o = 4.0 \ m/s$. As time changes the vertical speed of the jumper decreases linearly because of a constant downward gravitational acceleration. At the peak of the jump, his or her vertical speed goes to zero. Vertical speed versus time plot for this jumper with assumptions and parameters given above is presented in Figure 5-1.

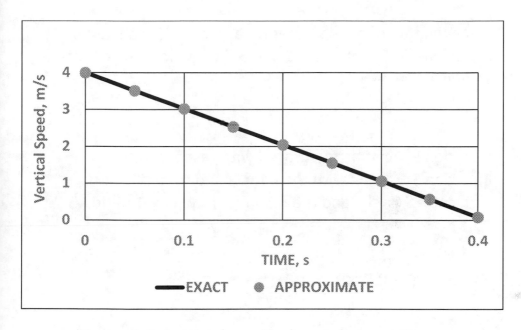

Figure 5-1: Jumper's vertical speed versus time

The jumper reaches his or her peak height in about 0.4 seconds and afterwards starts to descend. Approximate results in Figure 5-1 are obtained using Equation 5-1. Exact results in Figure 5-1 are obtained from Equation 5-6 by integrating Equation 5-2 by using the initial condition $at\ t = 0,\ S = S_o$ and the result is shown in Equation 5-6.

$$S = S_0 - g \times t \qquad\qquad\qquad 5\text{-}6$$

Since the rate of change of vertical speed with changing time is a constant, Equations 5-1 and 5-6 provide same results as shown in Figure 5-1.

Equation 5-4 can be integrated with respect to time using Equation 5-6 along with the initial condition $at\ t = 0,\ S = S_o$ and the result is shown in Equation 5-7.

$$Z = S_o \times t - 0.5 \times g \times t^2 \qquad\qquad 5\text{-}7$$

The change in the jumper's vertical position with respect to changing time can be obtained using either Equation 5-3 or Equation 5-7 as shown in Figure 5-2. For the present example, both equations give same results since the average speed during a time change of $(t_i - t_{i-1})$ is exactly $\frac{(S_i + S_{i-1})}{2}$

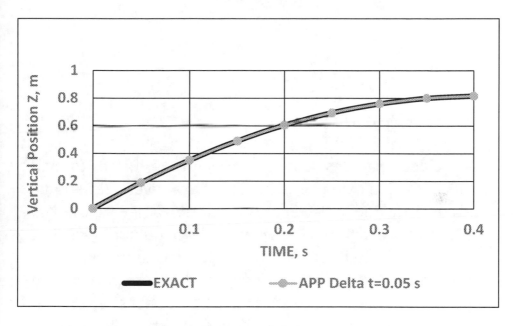

Figure 5-2: Jumper's vertical position Z versus time

Maximum kinetic energy of the jumper is at
$t = 0 \text{ and } Z = 0$, namely $0.5 \times M \times V_o^2$ which is 400 N-m.
At this point the jumper's potential energy, $M \times g \times Z$, is
zero. At the peak of his or her jump, his or her potential
energy reaches 400 N-m at a height of 0.815 meter and his or
her kinetic energy becomes zero. However, during the jump,
the sum of the potential and the kinetic energies is always
400 N-m.

Chapter 6

Vehicle Braking to a Stop

Vehicle decelerations depend heavily on tire surface to road surface traction characteristics. There are several different and sophisticated anti-skid braking systems that are used in vehicles today that prevent wheels to lock up during braking on any surface and improve vehicle control during braking. In this chapter we will simply investigate the stopping distance for a vehicle which has constant deceleration, i.e. $-a$. We will neglect the air resistance acting on the traveling vehicle. We will also neglect changes that are occurring in braking system components including tires during braking. Then the change in speed of a vehicle braking to a stop will be a linear one from the initial speed of S_i down to zero. Changing speed with respect to changing time from initial zero time to a final time t_f when the vehicle comes to a stop can be expressed in Equation 6-1.

$$\frac{0-S_i}{t_f-0} = -a \qquad\qquad 6\text{-}1$$

Since the vehicle's speed is decreasing linearly from its initial speed down to zero in Equation 6-1, its stopping distance, D_f , with changing time can be expressed as given in Equation 6-2.

$$\frac{D_f - 0}{t_f - 0} = \frac{(0 + S_i)}{2}$$

6-2

By combining Equations 6-1 and 6-2, the stopping distance, D_f , with changing initial speed can be expressed as shown in Equation 6-3.

$$D_f = \frac{S_i^2}{2 \times a}$$

6-3

When the vehicle is braking to a full stop, its kinetic energy is dissipated down to zero. This kinetic energy dissipation is equal to the work done during braking between the vehicle tires and the road's surface as shown in Equation 6-4.

$$0.5 \times m \times S_i^2 = m \times g \times \mu \times D_f$$

6-4

where m is vehicle's mass in kilograms, g is the gravitational acceleration on Earth's surface at 9.81 m/s², and μ is the kinematic coefficient of friction between the tires and the road's surface. By combining Equation 6-3 and 6-4, it can be shown that kinematic coefficient of friction between the tires and the road's surface is the ratio of a to g for this constant deceleration case.

Vehicle's stopping distance versus its initial speed when braking starts can be determined for different kinematic coefficients of friction between the tires and the road's surface using Equation 6-4 and the results are shown in Figure 6-1. μ is about 0.7 with new tires on a clean asphalt road. μ decreases to about 0.3 when the asphalt road is wet and slick. μ further decreases down to about 0.1, when the road is covered with ice. If we are traveling on an icy road at 100 km/hr, it will take us 400 meters to come to a full stop. Our constant deceleration will be 0.1xg or 0.981 m/s².

As the time interval gets small, the change in vehicle's speed with respect to changing time, Equation 6-1 can be rewritten in the following derivative form in Equation 6-5.

$$\frac{dS}{dt} = -a \qquad \text{6-5}$$

With the initial condition of $S = S_i$ at $t = 0$. Equation 6-5 can be integrated with respect to time to obtain the instantaneous speed of the vehicle as a linear function of time. The result is shown in Equation 6-6.

$$S = S_i - a \times t \qquad \text{6-6}$$

Since instantaneous speed is the derivative of instantaneous distance with respect to time, i.e. $\frac{dD}{dt} = S$, Equation 6-6 can be integrated again with respect to time using the initial condition $D = 0$ at $t = 0$ in order to obtain the instantaneous distance of the vehicle which is a parabolic function of time as shown in Equation 6-7.

$$D = S_i \times t - 0.5 \times a \times t^2 \qquad \text{6-7}$$

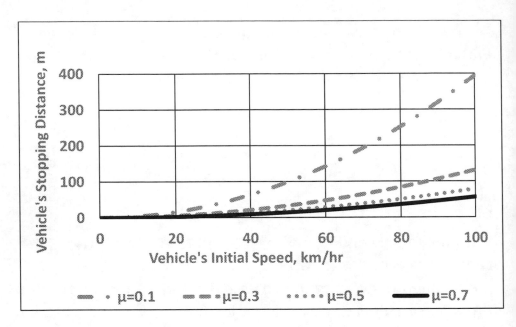

Figure 6-1: Vehicle's stopping distance versus its initial speed
for different kinematic coefficients of friction

Vehicle's linearly decreasing speed during braking versus
time is determined by using Equation 6-1 or Equation 6-6 for
S_i= 100 km/hr at t_i= 0 and for μ= 0.5 or a = 4.905 m/s². The
results are shown in Figure 6-2. The vehicle comes to a stop
in 5.66 seconds under given conditions.

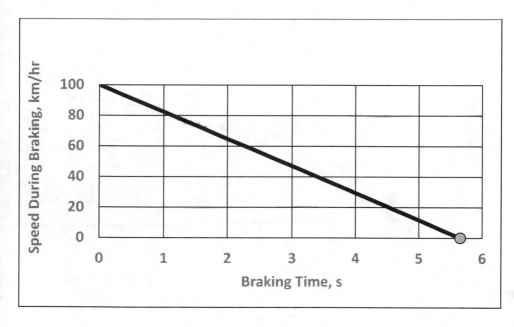

Figure 6-2: Vehicle's speed during braking versus time

for S_i = 100 km/hr and μ = 0.5

For conditions given for Figure 6-2 calculations, we can also calculate the distance traveled versus time by the vehicle to a full stop using Equation 6-2 or Equation 6-7. The behavior of the distance-time curve is a parabola for the present braking model. The results are shown in Figure 6-3. The vehicle comes to a stop at a distance of 78.65 meters in 5.66 meters.

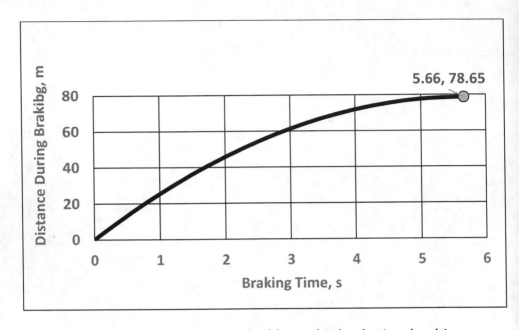

Figure 6-3: Distance traveled by vehicle during braking versus time for S_i = 100 km/hr and μ = 0.5

Chapter 7

Jumping Up Against Different Gravitational Acceleration Levels

We like to jump high on Earth's surface while participating in many sports such as basketball, track and field, ultimate frisbee, etc. against gravitational acceleration which is pulling us towards the center of the Earth. We do not notice changes in gravitational acceleration on the surface of our Earth because we are so far away from its center, i.e. about 6,300 kilometers. In the present jumping analysis, we will neglect drag resistance due to air in Earth's atmosphere. We wlll compare jumping speeds and hang times for a jumper on Earth's surface to a jumper on Jupiter and on our Moon.

Initially, the jumper is airborne upward from Earth's surface at $Z = 0$ with a speed of S_o at time $t = 0$. The only force acting on the jumper is the downward gravitational force. When the jumper is airborne, his or her vertical speed will change with respect to changing time as given in Equation 7-1.

$$\frac{S_i - S_{i-1}}{t_i - t_{i-1}} = -g \qquad \text{7-1}$$

where $g = 9.81 \, m/s^2$ is the gravitational acceleration on the surface of Earth, where $g = 22.9 \, m/s^2$ is the gravitational acceleration on the surface of Saturn and where $g = 1.67 \, m/s^2$ is the gravitational acceleration on the surface of our Moon. As the time interval $(t_i - t_{i-1})$ gets smaller and approaches zero, we can rewrite Equation 7-1 for instantaneous speed of the jumper as the first derivative of vertical speed with respect to time as given in Equation 7-2.

$$\frac{dS}{dt} = -g \qquad \text{7-2}$$

Equation 7-2 can be integrated with respect to time and the following Equation 7-3 is obtained by using the initial condition $S = S_o \, at \, t = 0$. The result is a linear relationship between jumper's speed and time.

$$S = S_o - g \times t \qquad \text{7-3}$$

As an example, if the jumper's initial upward speed is 5 meters per second, then it will take him or her to peak, i.e. $S = 0$, at 0.51 second on Earth's surface, 0.22 second on Jupiter's surface and 3.0 seconds on the surface of our Moon, namely $t_{peak} = \frac{S_0}{g}$. After peaking, the jumper will return to the surface at a downward speed which is the same as the initial upward speed. Total hang time for the jumper will be double the time to reach the peak, namely 1.02 seconds on Earth's surface, 0.44 second on Jupiter's surface and 6.0 seconds on the surface of our Moon, namely $t_{hang} = \frac{2 \times S_0}{g}$. Jumping speed versus time on Earth's surface, on Saturn's surface and on our Moon's surface are shown in Figure 7-1 for present example.

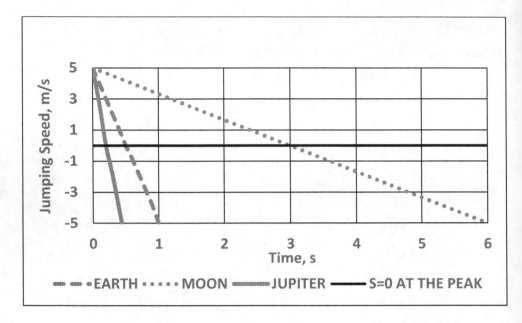

Figure 7-1: Speed versus time for a jumper
with an initial upward speed of 5 m/s

The change in the jumper's vertical position with respect to
changing time can be obtained from the jumper's vertical
speed-time profile with the following Equation 7-4.

$$\frac{Z_i - Z_{i-1}}{t_i - t_{i-1}} = \bar{S} = \frac{S_i + S_{i-1}}{2}$$ 7-4

where $\overline{S}$ is the jumper's average vertical speed during the time interval $(t_i - t_{i-1})$. As the time interval $(t_i - t_{i-1})$ gets smaller and approaches zero, we can rewrite Equation 7-4 for the instantaneous vertical position of the jumper. The first derivative of vertical position with respect to time using Equation 7-3 is shown in Equation 7-5.

$$\frac{dZ}{dt} = S = S_o - g \times t \qquad\qquad 7\text{-}5$$

Equation 7-5 can be integrated with respect to time and the following Equation 7-6 is obtained by using the initial condition $Z = 0 \; at \; t = 0$. The result is a parabolic relationship between jumper's instantaneous vertical position and time.

$$Z = S_o \times t - 0.5 \times g \times t^2 \qquad\qquad 7\text{-}6$$

As an example, using a jumper's initial upward speed of 5 meters per second, his or her vertical position versus time can be obtained using either Equation 7-4 or Equation 7-6. Results are shown in Figure 7-2 for Earth, Jupiter and our Moon.

With an initial upward speed of 5 meters per second, the jumper will reach a maximum vertical position of 1.27 meters on Earth's surface, 0.55 meter on Saturn's surface and 7.49 meters on our Moon's surface, namely a maximum vertical position of $Z_{max} = 0.5 \times \frac{S_o^2}{g}$.

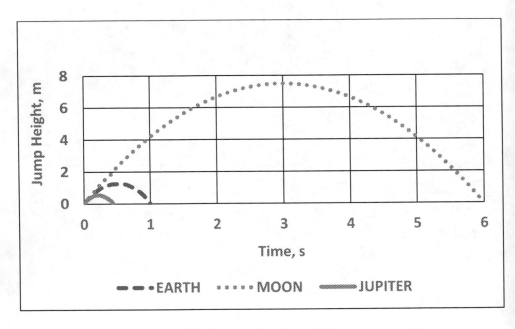

Figure 7-2: Jumper's vertical position versus time for initial upward speed of 5 m/s

Chapter 8

Capacitor Discharge

Let us investigate discharging of a fully charged capacitor which is connected to a resistor in a RC circuit in which there is only a resistor and a fully charged capacitor. Potential difference across the capacitor will decrease by time which will cause also cause a decrease in the absolute value of the potential difference across the resistor by time as given in Equation 8-1.

$$V_C + V_R = 0 \qquad\qquad 8\text{-}1$$

where V_C is the potential difference between the plates of the capacitor in volts and V_R is the potential difference across the resistor in volts. The potential difference between the plates of the capacitor is related to the charge, Q , in the capacitor in coulombs by a proportionality constant, C , in farads as shown in Equation 8-2.

$$Q = C \times V_C \qquad\qquad\qquad\qquad \text{8-2}$$

The potential difference across the resistor can be expressed in terms of the current i in amperes going through the resistor times the resistance R of the resistor in ohms. See Equation 8-3.

$$V_R = i \times R \qquad\qquad\qquad\qquad \text{8-3}$$

Change in charge with respect to change in time is defined as current, namely coulomb per second is an ampere, as shown in Equation 8-4.

$$i = \frac{dQ}{dt} \qquad\qquad\qquad\qquad \text{8-4}$$

Combining Equations 8-1 through 8-4, we can obtain the following Equation 8-5 in first order differential form for a discharging capacitor which is connected to a resistor in a RC circuit.

$$R \times \frac{dQ}{dt} + \frac{Q}{C} = 0 \qquad\qquad 8\text{-}5$$

The general solution to Equation 8-5 can be obtained by separating the variables Q and t which is given in Equation 8-6.

$$Q = Constant \times \exp(-\frac{t}{RC}) \qquad\qquad 8\text{-}6$$

where the "Constant" is determined from the fully charged capacitor's voltage, namely $Q = C \times V_{C\ initial}$ at $t = 0$. Potential difference across the capacitor will decrease by time as given below in Equation 8-7

$$V_C = \frac{Q}{C} = V_{C\ initial} \times \exp(-\frac{t}{RC}) \qquad\qquad 8\text{-}7$$

RC is the capacitor's discharging time constant in seconds which is the duration when the capacitor's charge reduces to 36.8% of its initial value.

Using Equations 8-2 and 8-5, a change in capacitor's potential difference with respect to changing time, $(t_{j+1} - t_j)$ can also be written as a first order difference relationship, see Equation 8-8.

$$V_{C\,j+1} - V_{C\,j} = V_{C\,j} \times (t_{j+1} - t_j)/RC \qquad\qquad 8\text{-}8$$

As an example, let us investigate discharging of a fully charged capacitor which is connected to a resistor in an RC circuit where

$V_{C\,initial} = 20\ volts,$

$C = 2x10^{-6}\ farads\ and$

$R = 5{,}000\ ohms\ .$

Capacitor's discharge, i.e. potential difference, versus time using the exact solution in Equation 8-7 and using the approximate solution (first order difference relationship) in Equation 8-8 are shown in Figure 8-1. For the approximate solution, a time increment of 0.001 second is used since the time constant for this example is 0.01 second.

The approximate solution gives us results that are very close to the exact solution. While using the first order difference relationship, time increments in our calculations should be at most 10% of the RC time constant value. Capacitor's discharge falls exponentially to zero.

The discharge is governed by the capacitor's discharging time constant RC. Large time constant means slow discharging capacitor and small time constant means fast discharging capacitor in a RC circuit.

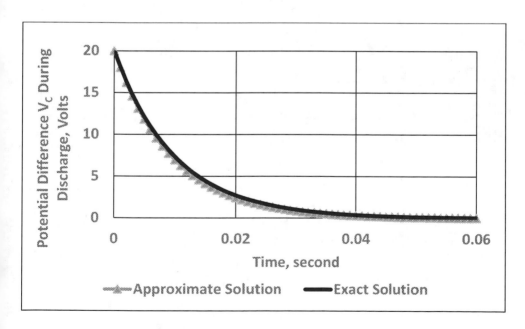

Figure 8-1: Capacitor discharge process in a RC circuit: Potential difference across the capacitor versus time

While the capacitor is discharging, the potential difference across the resistor can be obtained from Equation 8-1. Magnitudes of the potential difference across the resistor and therefore the current going through the resistor during the capacitor's discharge process are negative and they approach zero exponentially as shown in Figure 8-2.

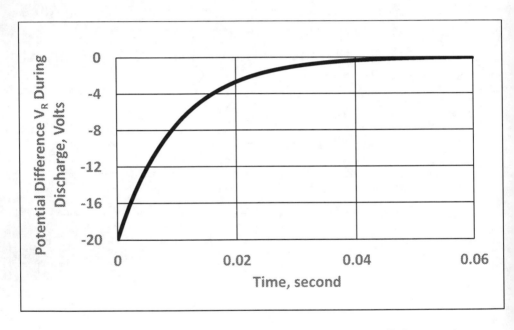

Figure 8-2: Capacitor discharge process in a RC circuit: Potential difference across the resistor versus time

Chapter 9

Spread of Contagious Diseases

In this chapter, we will investigate the spread of a contagious disease in an isolated town. Outside effects due to travel and mixing of town's people with other region's people are not allowed in the present analysis. A simple model to predict the spread of a contagious disease can be formulated, if we know the number of people who are infected by the disease at a given time in the town and the population of the town at the same time. Number of people in town who have the contagious disease increases with changing time as a product of the number of people who have the disease and the number of people who do not have the disease at a given time.

Let us denote the number of people in town who are infected by a contagious disease by C and the population of the town by P. An increase in C with respect to changing time, i.e. $(t_{i+1} - t_i)$, is given in Equation 9-1 as a second order difference relationship.

$$\frac{C_{i+1}-C_i}{t_{i+1}-t_i} = r \times C_i \times (P - C_i)$$

9-1

where r is the disease spreading constant. For the present example, we will obtain r by assuming that 2 more people get infected by the contagious disease one week after it was diagnosed on 10 people in town with a population of 20,000. Using these initial conditions, r can be obtained using Equation 9-1 and the result is shown in Equation 9-2.

$$r = \frac{(12-10)}{(1-0)} \times \frac{1}{10 \times (20,000-10)} = 1.0005 \times 10^{-5} \quad \frac{1}{week \times \# \ of \ people}$$

9-2

Using Equations 9-1 and 9-2, number of contagious disease infected people in this isolated town can be determined versus time in weeks as shown in Figure 9-1. 50% of the town's people will be infected in 41 weeks. 99% of town's people will be infected in 63 weeks.

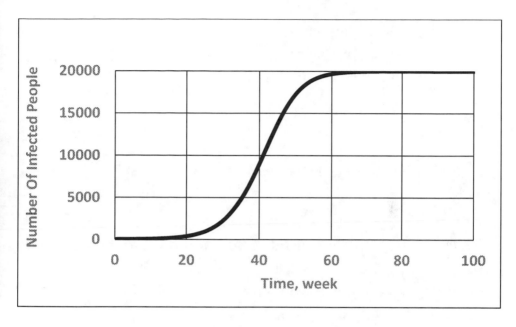

Figure 9-1: Disease infected people versus time
using second order difference relationship

As time change gets smaller, we can rewrite Equation 9-1 as a
first order non-linear differential equation as given in Equation
9-3.

$$\frac{dC}{dt} = r \times C \times (P - C)$$ 9-3

By separating variables C and t , Equation 9-3 can be integrated to obtain the following Equation 9-4 for spreading of a contagious disease in an isolated town.

$$C = \frac{1}{1+exp[-P\times(r\times t+CONS)]} \qquad \text{9-4}$$

$CONS$ is the constant of integration and it can be obtained from the initial conditions at $t = 0$, namely 10 people are diagnosed with a contagious disease in an isolated town with a population of 20,000. Then

$CONS = -0.00038 \; \frac{1}{\# \, of \, people}$. Equation 9-4 takes the following form.

$$C = \frac{1}{1+exp[-P\times(r\times t-0.00038)]} \qquad \text{9-5}$$

Infected people versus time is determined using the approximate second order difference relationship in Equation 9-1. One week time intervals are used while calculating the approximate results. The approximate results are compared with the exact solution to the first order non-linear differential Equation 9-5.

Comparative results are shown in Figure 9-2. Exact solution to the present example shows us that 50% of the town's people will be infected in 42 weeks. 99% of town's people will be infected in 67 weeks. Approximate solution results are very close to the exact solution results as depicted in Figure 9-2.

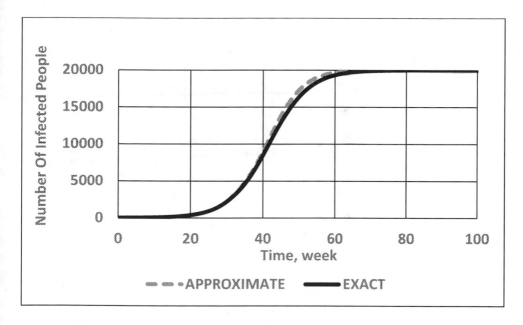

Figure 9-2: Comparison of approximate solution to exact solution for infected people versus time

As more people gets infected by the contagious disease one week after it was diagnosed on 10 people, the disease spreading constant r increases, and therefore time for the disease to spread to the whole town gets shorter and shorter as shown in Figure 9-3.

If 6 people more get infected after the first week that the contagious disease is diagnosed, 50% percent of town can get infected in 16 weeks and 99% of town can get infected in 26 weeks.

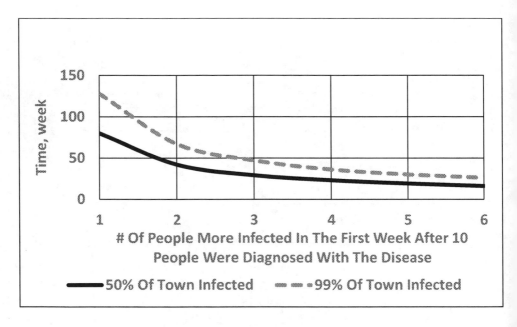

Figure 9-3: Time versus more # of people that are infected in the first week after 10 people were diagnosed with the disease

Chapter 10

Evaporation of a Raindrop

A raindrop's evaporation during its descend to the ground can be a very complicated physical phenomenon. Its shape, its temperature, surrounding's temperature, pressure, relative humidity and contaminations, wind conditions, solar radiation, and its impact with other raindrops can affect a raindrop's evaporation phenomenon. Let us assume a spherical raindrop which evaporates in proportion to its surface area. Also let us assume that after several studies and observations, a constant proportionality constant can be determined. So a change in the volume of a raindrop with respect to changing time can be approximated by the following finite difference Equation 10-1.

$$\frac{\frac{4}{3} \times \pi \times \left(R_{i+1}^3 - R_i^3\right)}{(t_{i+1} - t_i)} = -C \times 4 \times \pi \times R_i^2 \qquad \text{10-1}$$

where R_i is the radius of the raindrop in millimeters at time t_i in minutes and R_{i+1} is the radius of the raindrop at time t_{i+1}. C is the proportionality constant in mm/min.

The minus sign in front of C comes from the fact that the radius of the raindrop decreases by time as the raindrop evaporates. Equation 10-1 can be simplified to determine R_{i+1} as shown in Equation 10-2.

$$R_{i+1} = \left[R_i^3 - 3 \times C \times R_i^2 \times (t_{i+1} - t_i)\right]^{1/3} \qquad \text{10-2}$$

Let us do an example by using Equation 10-2 along with a proportionality constant of 0.2 mm/min for a raindrop which initially has a radius of 2 mm. Under these conditions, this raindrop will evaporate in 10 minutes as shown in Figure 10-1. Also the radius of the raindrop decreases linearly with time. It is easy to show (later in this chapter) this linear relationship from the differential form of Equation 10-1. So a change in the radius of a raindrop with respect to changing time can be determined by the following linear finite difference Equation 10-3.

$$R_{i+1} = R_i - C \times (t_{i+1} - t_i) \qquad \text{10-3}$$

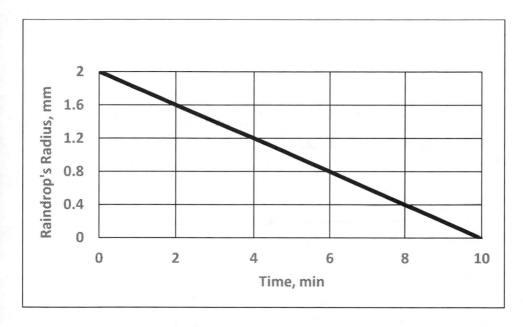

Figure 10-1: A raindrop's radius which is initially at 2 mm versus time during evaporation

Equation 10-1 can be written in the following differential form as shown in   Equation 10-4.

$$\frac{d\left(\frac{4}{3} \times \pi \times R^3\right)}{dt} = -C \times 4 \times \pi \times R^2 \qquad \text{10-4}$$

However, $dR^3 = 3 \times R^2 \times dR$ and Equation 10-4 becomes the following first order differential Equation 10-5.

63

$$\frac{dR}{dt} = -C$$
10-5

The solution to Equation 10-5 is given in Equation 10-6 in which R_o is the initial radius of the raindrop.

$$R = R_o - C \times t$$
10-6

Equation 10-6 shows that the radius of a raindrop decreases linearly with respect to time. Same results can be obtained by using the linear finite difference Equation 10-3. Total evaporation times for different size raindrops are obtained from Equation 10-6 as $R \to 0$. Results that are shown in Figure 10-2 use a proportionality constant of 0.2 mm/min.

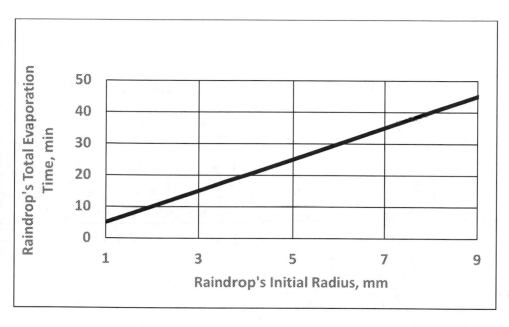

Figure 10-2 Total evaporation time for raindrops
with different initial radll

Chapter 11

CO Release from a Kerosene Heater

Carbon monoxide, CO , release from a kerosene heater can be very dangerous for people and animals in an unventilated room. If you operate a kerosene heater in an unventilated room, you can get CO poisoning which can be very unhealthy for humans and for animals above 35 PPM (parts per million) of CO in room's air. Maximum recommended average safe limit for carbon monoxide in a room is 9 PPM. Let us investigate the change in mass of carbon monoxide in an unventilated room with changing time.

First let us define the PPM of CO in the room as shown in Equation 11-1.

$$PPM\ CO = \left(\frac{Mass\ of\ CO\ in\ the\ room}{Mass\ of\ AIR\ in\ the\ room}\right) \times 10^6 \qquad 11\text{-}1$$

In our present analysis, we will assume a constant mass of air in the room. The unventilated room is assumed to have a volume of 40 m³and the density of air in the room at 20 °C is assumed to be 1.194 kg/m³, then we have 47.76 kg of air in this room, namely $m_{air} = 47.76\ kg$.

Let us also assume that our kerosene heater releases 10 mg of CO to the room every minute it is on, namely $\dot{m}_{CO} = 10\ mg/min$. With the above information, we can calculate the increase in the carbon monoxide level in the room with respect to changing time using the following finite difference Equation 11-2.

$$\frac{PPM\ CO_{i+1} - PPM\ CO_i}{t_{i+1} - t_i} = CONSTANT = \frac{\dot{m}_{CO}}{m_{air}} \times 10^6 \qquad 11\text{-}2$$

The $CONSTANT$ in Equation 11-2 has the dimension $PPM\ CO$ per minute. Carbon monoxide level in an unventilated room changes linearly with changing time. Carbon monoxide level increase in an unvented room when a kerosene heater is on is determined using the above parameters and Equation 11-2.

The results are shown in Figure 11-1. The CO level in this unventilated room increases linearly with respect to time and it reaches its maximum recommended average safe limit of 9 PPM in 43 minutes.

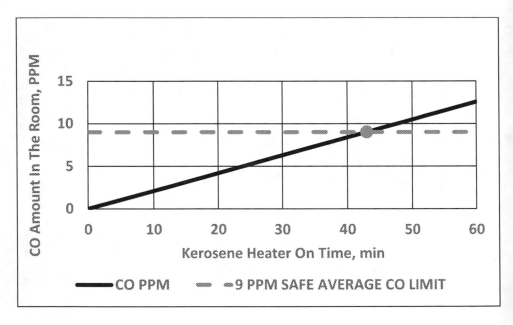

Figure 11-1: Carbon monoxide level increase versus time in an unventilated room

The finite difference Equation 11-2 can be written in differential form as shown in Equation 11-3.

$$\frac{d(PPM\ CO)}{dt} = CONSTANT \qquad\qquad 11\text{-}3$$

Equation 11-3 is integrated easily by using the initial condition $PPM\ CO = 0$ at $t = 0$. The linear relationship between the carbon monoxide levels in the unventilated room with respect to time is given in Equation 11-4.

$$PPM\ CO = CONSTANT \times t \qquad\qquad 11\text{-}4$$

We can use either Equation 11-2 or 11-4 to determine CO levels in the room versus time, if we know the kerosene heater's CO release rate, the unventilated room's volume and the density of air in the room. Both equations will give us the same result since the mass of carbon monoxide in an unventilated room changes linearly with time. Figure 11-2 shows CO levels increase in a 40 m³ unventilated room in which the air mass is 47.76 kg for different kerosene heaters' CO release rates.

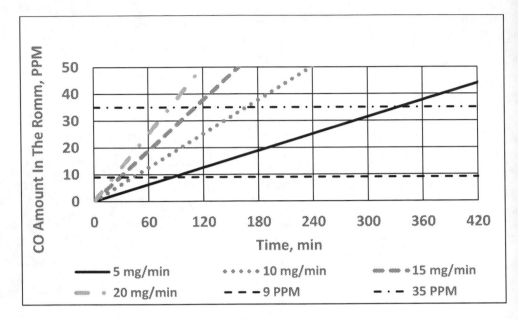

Figure11-2: CO level increase in an unventilated room versus time for different CO release rates

Time it takes to reach 9 PPM and 35 PPM CO levels in the unventilated room for different kerosene heaters' CO release rates are presented in Table 11-1.

Kerosene Heater CO Release Rate, mg/min	Time To Reach 9 PPM CO Level, min	Time To Reach 35 PPM CO Level, min
5	21.5	83.6
10	28.7	111.4
15	43.0	167.2
20	86.0	334.3

Table 11-1: Time it takes to reach 9 PPM and 35 PPM CO levels in a 40 m³ unvented room for different CO release rates

Chapter 12

Population Growth or Decline

Changes in population of people and animals with respect to changing time can be predicted by product of a constant population change rate, R , and the population present at the initial time. This type of analysis excludes sudden occurrences in population change due to natural disasters, epidemics, migrations, wars, etc. Changes in population with respect to changing time can be written in finite difference form as shown in Equation 12-1.

$$\frac{P_{i+1}-P_i}{t_{i+1}-t_i} = R \times P_i \qquad\qquad\qquad 12\text{-}1$$

where $(P_{i+1} - P_i)$ is the population increase or decrease during an increase in time, namely $(t_{i+1} - t_i)$. For increasing population, the constant population change rate is positive. For example if the population is increasing 2% per year, then $R = 0.02\ per\ year$. For decreasing population, the constant population change rate is negative. For example if the population is decreasing 5% per year, then $R = -0.05\ per\ year$.

Let us do an example for a city's population increase projections using Equation 12-1. Initially the city has a population of 20,000. Past population change data and all future economic forecasts show that this city's population can grow a minimum of 2% a year and a maximum of 4% a year. Population increase versus time for this city is shown in Figure 12-1. If the constant population change rate is 2% at a minimum, the city's population will double in 34.7 years. If the constant population change rate is 4% at a maximum, the city's population will double in 17.3 years. For the average constant population change rate of 3%, the city's population will double in 23.1 years.

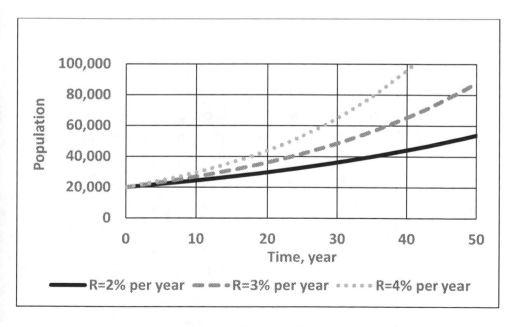

Figure 12-1: Population increase versus time

Equation 12-1 can be written in differential form as the change in time goes to zero which is shown in Equation 12-2.

$$\frac{dP}{dt} = R \times P \qquad\qquad 12\text{-}2$$

The exact solution to Equation 12-2 is the exponential function. If an initial condition of $P = P_o$ is used, the solution to Equation 12-2 takes the following form in Equation 12-3.

$$P = P_o \times e^{R \times t} \qquad\qquad 12\text{-}3$$

We can use either Equation 12-1 or Equation 12-3 in calculating population growth or decline as long as we know values for P_o and R . However, as the population change rate gets large, the finite difference equation overestimates the exact solutions obtained from Equation 12-3. For large population change rates, smaller time intervals should be used when Equation 12-1 is utilized.

Let us analyze another case. This time let us look at the mosquito population decline in a fresh water pond. Biologists estimate that mosquito decline rate will be -10% per year when mosquitos' larvae growth is controlled by using effective larvicide pellets. If we use mosquito eating fish in addition to larvicide pellets, the mosquito decline rate will increase to -20% per year. However, if we circulate all the water in the pond and leave no shallow and stagnant regions, the mosquito decline rate will increase to -50% per year. Using Equation 12-1 and one year time intervals, mosquito population decrease versus time for the fresh water pond for three different population decline rates is shown in Figure 12-2.

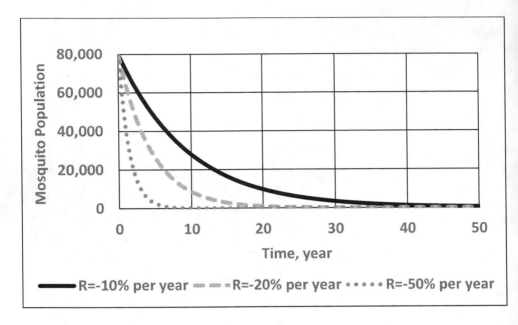

Figure12-2: Mosquito population decrease versus time for three different population decline rates

For -10% decline rate, it takes 43.7 years to get rid of 99% of mosquitos. For -20% decline rate, it takes 20.7 years to get rid of 99% of mosquitos. For -50% decline rate, it takes 6.7 years to get rid of 99% of mosquitos in this fresh water pond.

Results in Figure 12-2 was obtained using Equation 12-1 with one year time intervals.

This approximate evaluation overestimates the decline of the population when there is a large population change rate as shown in Figure 12-3 for $R = -0.2$. Similar overestimation is true for population growths with large change rates. If we still want to use Equation 12-1 for a large change rate, we have to reduce the time interval used during our calculations.

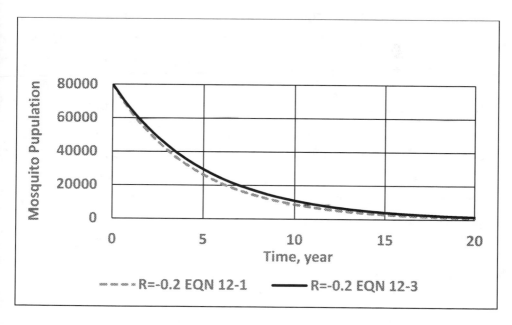

Figure 12-3: Approximate and exact solutions to mosquito population decline for $R = -0.2$

Chapter 13

Pressure Change in the Atmosphere

As we go higher in altitude in our Earth's atmosphere, the density of air decreases so does the air pressure forces acting on our bodies. Changes in pressure as a result of changes in altitude can be modeled by balancing out the pressure and gravitational forces acting on a body. Let us consider a thin element with a thickness dz in our atmosphere that is perpendicular to the radial direction z which is outward from the center of the Earth towards outer space. Pressure and gravitational forces acting on this thin element balance as shown in Equation 13-1.

$$P \times A - (P + dP)A - \rho \times g \times A \times dz = 0 \qquad \text{13-1}$$

$P \times A$ is the pressure force, P , in newton per m^2 (pascal) acting on the bottom surface area, A , in m^2 of the thin element. $(P + dP) \times A$ is the pressure force acting on the top surface area of the thin element. So the net pressure force acting on this thin element is $-dP \times A$ where dP is the pressure change between the bottom and the top surfaces of the thin element.

The gravitational force acting on the thin element is $-\rho \times g \times A \times dz$ where ρ is the atmospheric air density in kg/m³, g is the gravitational acceleration towards the center of the Earth in m/s² and dz is the thickness of the thin element in meters and it is positive in the increasing altitude direction. Equation 13-1 can be reduced to Equation 13-2 as follows.

$$\frac{dP}{dz} = -\rho \times g \qquad \text{13-2}$$

If the atmospheric air is assumed to behave like an ideal gas in a constant temperature and constant gravitational acceleration environment, air pressure and air density are related as shown in Equation 13-3.

$$\frac{P}{\rho} = \frac{P_{atm}}{\rho_{atm}} \qquad \text{13-3}$$

where P_{atm} is the sea level air pressure at 101,325 N/m² and ρ_{atm} is the sea level air density at 1.21 kg/m³ .Also g at sea level is 9.81 m/s² and it is assumed to be a constant at elevations up to 10 km for the present analysis.

Combining Equations 13-2 and 13-3 provide us the following finite difference Equation 13-4 for the pressure drop in our Earth's atmosphere.

$$\frac{P_{i+1}-P_i}{z_{i+1}-z_i} = -\left(\frac{\rho_{atm} \times g}{P_{atm}}\right) \times P_i \qquad \text{13-4}$$

In Equation 13-4, the constant $\left(\frac{\rho_{atm} \times g}{P_{atm}}\right)$ has a value of $0.000117 \frac{1}{m}$ or $0.117 \frac{1}{km}$. As $(z_{i+1} - z_i)$ approaches zero, we obtain the differential form of Equation 13-4 as shown in in Equation 13-5.

$$\frac{dP}{dz} = -\left(\frac{\rho_{atm} \times g}{P_{atm}}\right) \times P \qquad \text{13-5}$$

Equation 13-5 can be integrated by separating variables and using the sea level as the reference, namely $P = P_{atm}$ at $z = 0$ .Then we obtain an exact solution for pressure P at an altitude h as shown in Equation 13-6.

$$P = P_{atm} \times exp\left[-\left(\frac{\rho_{atm} \times g}{P_{atm}}\right) \times h\right]$$ 13-6

Atmospheric air pressure drop as a function of altitude can be calculated using either Equation 13-4 or Equation 13-6. The results are shown in Figure 13-1. Half a kilometer height intervals are used during atmospheric air pressure drop calculations using Equation 13-4. Both equations give very similar results. Approximate solutions obtained using the finite difference Equation 13-4 starts to deviate from the exact solution of Equation 13-6 as the altitude increases. Equation 13-4 predicts a 3.5% less air pressure at 10 km altitude as compared to Equation 13-6. Difference in high altitude pressure values will decrease if we use small height intervals in our calculations when we use Equation 13-4.

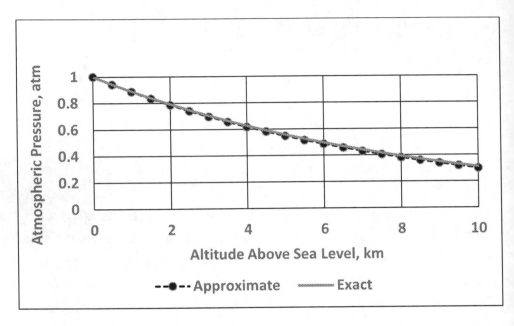

Figure13-1: Atmospheric air pressure drop
as a function of altitude above sea level

Atmospheric air pressure values in Figure 13-1 are very close to actual measured data up to 6 km of altitude. At higher altitudes the atmospheric air pressure measurements are lower, i.e. at 10 km altitude, atmospheric air pressure data is 0.26 atm versus the 0.31 atm value we obtain from Equation 13-6.

Chapter 14

Pressure Change under the Sea

When we dive into the sea, the pressure around us increases very fast due to high density of sea water which is about 1,030 kg/m³. Most divers should not go deeper than forty meters without special equipment where the sea water pressure reaches five atmospheres and then nitrogen narcosis sets in causing loss of sense and movement. Changes in pressure in water as a result of changes in depth can be modeled by balancing out the pressure and gravitational forces acting on a body similar to the analysis we have done in Chapter 13. Let us consider a thin element with a thickness dz in sea water where dz is perpendicular to the sea surface and z is positive in the increasing depth direction. Also sea water pressure increases above atmospheric pressure as z increases. Pressure and gravitational forces acting on this thin element balance as shown in Equation 14-1.

$$P \times A - (P + dP)A + \rho \times g \times A \times dz = 0 \qquad \text{14-1}$$

$P \times A$ is the pressure force, P , in newton per m² (pascal) acting on the top surface area, A in m², of the thin element. $(P + dP) \times A$ is the pressure force acting on the bottom surface area of the thin element. So the net pressure force acting on this thin element is $-dP \times A$ where dP is the pressure change (increase) between the top and the bottom surfaces of the thin element. Gravitational force acting on this thin element is $+\rho \times g \times A \times dz$ where $\rho = 1{,}030 \ kg/m^3$ is the sea water density and $g = 9.81 \ m/s^2$ is the gravitational acceleration towards the center of the Earth. In the present analysis, we will assume a constant sea water density while neglecting, salinity, temperature and depth effects on sea water density. We will also assume a constant gravitational acceleration. Equation 14-1 can be reduced to Equation 14-2 as follows:

$$\frac{dP}{dz} = \rho \times g \qquad\qquad 14\text{-}2$$

The increasing sea water pressure with increasing depth can be written in a finite difference form as shown in Equation 14-3.

$$P_{i+1} - P_i = \rho \times g \times (z_{i+1} - z_i) \qquad\qquad \text{14-3}$$

Equations 14-2 can also be integrated easily since the right hand side of Equation 14-2 is a constant. Integration from P_{atm} at $z = 0$, namely the atmospheric pressure at the sea surface, to P_D at a depth D , i.e. the sea water pressure at a certain depth, gives us the following Equation 14-4.

$$P_D - P_{atm} = \rho \times g \times D \qquad\qquad \text{14-4}$$

Equation 14-4 can be rearranged and written in a more convenient form, namely the ratio of sea water pressure at depth D to atmospheric pressure at zero depth, as shown in Equation 14-5.

$$\frac{P_D}{P_{atm}} = 1 + \left(\frac{\rho \times g}{P_{atm}}\right) \times D = 1 + 0.0997 \times D \qquad\qquad \text{14-5}$$

We can use in our calculations either the finite difference form of Equation 14-2 or the integrated form, since $\frac{dP}{dz} = constant$. We will obtain the same result. Linear variation of pressure with depth below sea level is presented in Figure 14-1. At a 10 meter depth, the sea water pressure doubles to the atmospheric sea level pressure.

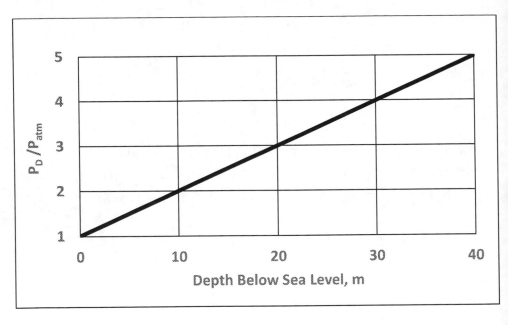

Figure 14-1: Change in pressure below sea level

The deepest sea level on our Earth is in the Pacific Ocean called the Marianas Trench which is 10,923 meters deep and the pressure at that level is 1,090 atmospheres.

Chapter 15

Radioactive Material's Decay

Changes in mass of a radioactive material with respect to changing time is a first order chemical reaction and can be predicted by the product of a constant decay rate, α , and the mass of the radioactive material present at initial time. The radioactive decay analysis that is presented here is similar to the analysis presented in Chapter 12 for population decrease. For example, radioactive carbon, C-14 , is found in living matter during its life, stays in living matter after its death such as in bones and in wood. C-14 disintegrates into a non-radioactive material slowly after death and it loses half of its mass, i.e. half-life, in 5,730 years. Ages of bones and wood can be estimated up to 60,000 years by determining the amount of radioactive carbon material left in them.

Changes in mass of radioactive material with respect to changing time can be written in finite difference form as shown in Equation 15-1.

$$\frac{R_{i+1}-R_i}{t_{i+1}-t_i} = -\alpha \times R_i \qquad\qquad 15\text{-}1$$

where $(R_{i+1} - R_i)$ is the decrease in mass of a radioactive material during an increase in an interval of time, namely $(t_{i+1} - t_i)$. The decay rate α is always positive and most commonly has the dimension 1/year. As $(t_{i+1} - t_i)$ goes to zero, Equation 15-1 takes the following first order differential equation form in Equation 15-2.

$$\frac{dR}{dt} = -\alpha \times R \qquad\qquad 15\text{-}2$$

Equation 15-2 can be integrated after separating the variables and by using the initial mass of the radioactive material R_o at $t = 0$ for determining the integration constant. The result is given in Equation 15-3.

$$\frac{R}{R_o} = e^{-\alpha \times t} \qquad\qquad 15\text{-}3$$

For example, radioactive carbon, C-14, has a half-life of 5,730 years. We can determine the decay rate for C-14 using Equation 15-3 as shown in Equation 15-4.

$$\alpha = -\left(\frac{1}{5,730}\right) \times \ln\left(\frac{1}{2}\right) = 0.000121 \quad 1/year \qquad\qquad 15\text{-}4$$

Using the decay rate in Equation 15-4, it can be shown that 90% of radioactive carbon, C-14, material's mass disintegrates in fossils in 19,035 years, 99% in 38,069 years and 99.9% in 57,104 years. This is the reason scientists limit radioactive dating method using C-14 for once lived matter to 60,000 years.

Let us do another example. Consider 10 mg of a radioactive material which has a half-life of 3 days. The decay rate for this material is calculated in Equation 15-5.

$$\alpha = -\left(\frac{1}{3}\right) \times \ln\left(\frac{1}{2}\right) = 0.231 \quad 1/day \qquad\qquad 15\text{-}5$$

Using the above decay rate and Equations 15-1 and 15-3, disintegration of this radioactive material with respect to time is shown in Figure 15-1. The finite difference Equation 15-1 underestimates the decaying process. In our calculations, when one-day time interval is used, half-life comes out to be 2.67 days, namely an 11.1% error. When the time interval for the calculations is reduced to 0.125 day, half-life calculation improves to 2.96 days, namely only a 1.4% error.

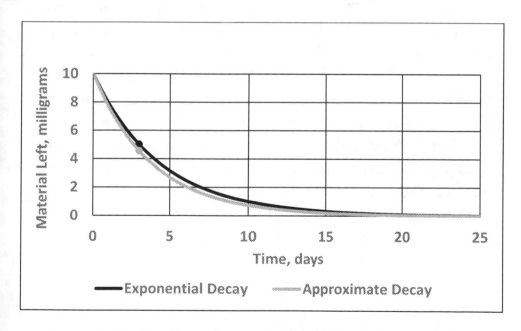

Figure 15-1: A radioactive material with a 3-day half-life: Decay versus time

Chapter 16

Water Flowing Out Of a Hole Centered at the Bottom of a Cylindrical Tank

Let us consider a large cylindrical water tank with a small drain hole centered at the bottom of it. Assume that the tank has a radius R in meters and the initial height of the water in the tank is H in meters. The small drain hole is a circular one with a radius r in meters. We will also assume a frictionless and a non-rotating flow of an incompressible fluid, i.e. water, throughout the tank and the drain hole. Then the governing fluid mechanics equations simplify to the Bernoulli's principle along a fluid's streamline. Velocity of the fluid particles exiting the discharge hole at the bottom center of the vessel is time dependent and it depends on the potential energy available between positions at the top of the water level in the tank and the drain hole at the bottom center of the tank. The jet stream that leaves the tank from the drain hole has losses due to friction and due to contraction of the jet stream. These losses are determined experimentally and specified as a dimensionless coefficient of discharge, i.e., C_d . The rate of volume flow that comes out of the drain hole can be defined by using the Bernoulli's principle and a dimensionless coefficient of discharge, C_d as shown in Equation 16-1.

$$Q_{out} = A_{drain} \times C_d \times \sqrt{2 \times g \times h} \qquad\qquad \text{16-1}$$

where Q_{out} is the volume flow rate from the drain hole in m³/s, $A_{drain} = \pi \times r^2$ is the area of the drain hole in m², $g = 9.81$ is the gravitational acceleration in m/s², and h is the height of the water in the cylindrical tank in meters at time t in seconds. Equation 16-1 considers conservation of mechanical energy along a streamline in a steady flow, namely no changes are occurring with respect to time in variables such as pressure, density and flow velocity along a streamline from top of the water in the tank to the drain hole at the bottom center of the tank. Then the water volume flow rate out of the drain hole should balance out water's volume decrease in the tank with changing time as shown in Equation 16-2.

$$-A_{tank} \times \frac{dh}{dt} + A_{drain} \times C_d \times \sqrt{2 \times g \times h} = 0 \qquad\qquad \text{16-2}$$

where $A_{tank} = \pi \times R^2$. Equation 16-2 can be written in a finite difference form that provides the change in water height in the water tank with changing time and it is given in Equation 16-3.

$$h_{i+1} = h_i - \left(\frac{A_{drain} \times C_d \times \sqrt{2 \times g}}{A_{tank}}\right) \times \sqrt{h_i} \times (t_{i+1} - t_i) \qquad \text{16-3}$$

Exact solution to Equation 16-2 can be obtained by separating the variables h and t and integrating it from the initial height of the water in the tank H to h at time t. The exact solution to Equation 16-2 is shown in Equation 16-4.

$$h = \left[\sqrt{H} - 0.5 \times \left(\frac{A_{drain} \times C_d \times \sqrt{2 \times g}}{A_{tank}}\right) \times t\right]^2 \qquad \text{16-4}$$

Equations 16-3 and 16-4 are used to calculate the changes in water height in the tank versus changing time. Following parameters are used for the present water height versus time calculations:

$$H = 2\ meters, \quad R = 1\ meter, \quad r = 0.01\ m \quad and \quad C_d = 0.95$$

Also 60 second time intervals are used for the finite difference equation's approximate solutions. Water levels in the tank versus time is shown in Figure 16-1.

The finite difference equation's solutions follow the exact solution very closely at high water levels in the tank. With advancing time, as the water level in the tank decreases, the finite difference solutions start to underestimate the water levels in the tank. This error increases over 5% as the water level reaches 0.1 meter. Difference between the exact and the approximate solutions for low water levels are shown in Figure 16-2. In order to minimize this low water level error, we have to take smaller time intervals while using the finite difference Equation 16-3.

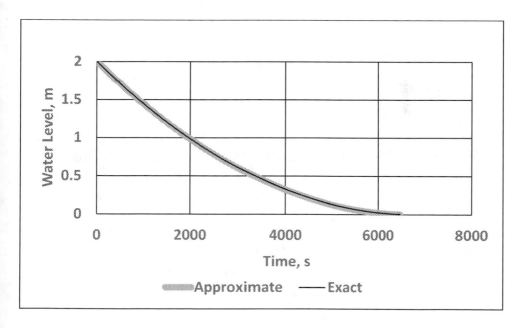

Figure 16-1: Water levels in the tank versus time

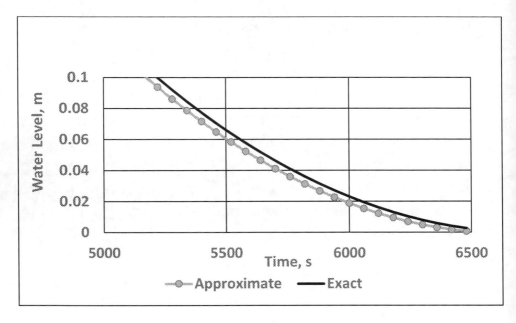

Figure 16-2: Difference between exact and approximate solutions for low water levels in the tank

Water levels in the tank versus time for various circular drain holes radii are shown in Figure 16-3. As the drain hole radius is increased, complete drainage time for the tank decreases very fast, namely 400 minute drainage time for a 0.005 meter drain hole radius versus 10 minute drainage time for a 0.03 meter drain hole radius, as shown in Figure 16-4.

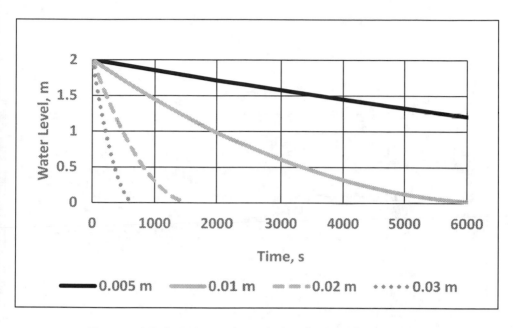

Figure 16-3: Water levels in the tank versus
time for various circular drain hole radii

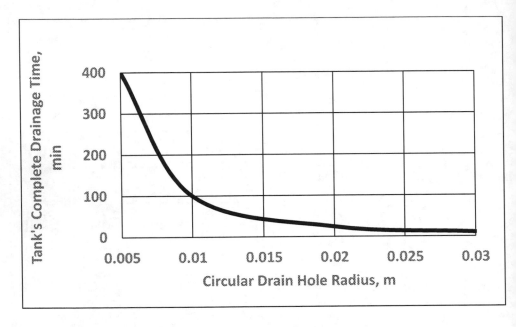

Figure 16-4: Tank's complete drainage times versus circular drain hole radius

Chapter 17

Bacteria Growth

Bacteria multiplies rapidly under right conditions and nutrients. The number of bacteria increases exponentially which can be modeled very similar to a continuously compounding interest. An increase in the number of bacteria per a change in time can be defined as the product of a bacteria population growth rate and the bacteria present at zero time. Finite difference representation of bacteria population growth is shown in Equation 17-1.

$$\frac{N_{i+1}-N_i}{t_{i+1}-t_i} = r \times N_i \qquad \text{17-1}$$

where $(N_{i+1} - N_i)$ is the increase in the number of bacteria during a change in time $(t_{i+1} - t_i)$ and r is the bacteria population growth rate with the dimension of per unit time. As $(t_{i+1} - t_i)$ goes to zero Equation 17-1 can be written in the differential form as shown in Equation 17-2.

$$\frac{dN}{dt} = r \times N \qquad\qquad\qquad 17\text{-}2$$

Equation 17-2 can be integrated after separating the variables N and t and applying the condition N_o for the initial number of bacteria. The exact solution to Equation 17-2 is provided in Equation 17-3.

$$\frac{N}{N_o} = e^{r \times t} \qquad\qquad\qquad 17\text{-}3$$

Let us do an example for a bacteria population that increases 9% per week. The bacteria population growth rate for this example can be calculated from Equation 17-3 as shown below.

$$r = \frac{\ln 1.09}{7} = 0.012311 \quad per\ day$$

Using the finite difference representation of bacteria population growth in Equation 17-1 and the exponential representation in Equation 17-3, bacteria population growth curves versus time are shown in Figure 17-1.

The bacteria population doubles in 56.3 days in this case. The approximate solution results in a 0.42% lower prediction of the bacteria population doubling time when one day time intervals are used during finite difference calculations.

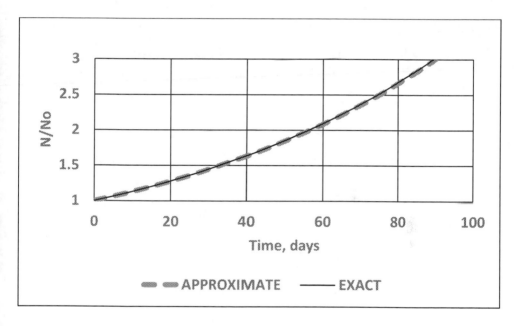

Figure 17-1: Bacteria population growth versus time

In another example, the bacteria population doubles every 20 minutes. The bacteria population growth rate for this example can be calculated from Equation 17-3 as given below.

$$r = \frac{\ln 2}{(20/60)} = 2.079442 \quad per\ hour$$

The bacteria population growth curves for a bacteria that doubles every 20 minutes are shown for Equations 17-1 and for 17-3 in Figure 17-2. After two hours, the bacteria population increases 64 fold. The finite difference approximation underestimates the population increase after two hours by 4.2%, namely as 61.3 fold. During finite difference calculations for this example, 0.01 hour time increments are used.

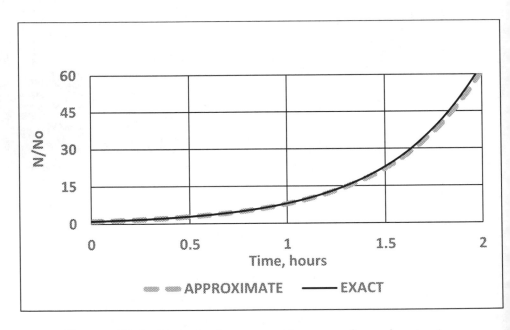

Figure 17-2: Population growth curve for a bacteria that doubles every 20 minutes

Chapter 18

Optimum Book Sale Price

Sales volume of a new book can be very price sensitive. A publishing company is going to release a new book and they want to price the book in such a way that they can maximize their revenues from its sales. The marketing department of the publishing company does a survey and a statistical analysis to determine the optimum price for this new book.

Results of the marketing study for this new book predict that 60,000 books will be sold at a $5.00 price. Also a $0.50 decrease from the $5.00 unit price will increase this book's sales on the average 10,000 units within 95% confidence levels. The upper limit of the 95% confidence level predicts that a $0.50 decrease from the $5.00 unit price will increase the book's sales only by 11,000 units. The lower limit of the 95% confidence level predicts that a $0.50 decrease from the $5.00 unit price will increase the book's sales only by 9,000 units.

For the upper 95% confidence limit, the number of books, N , that can be sold versus unit price, P , is shown in Equation 18-1.

$$N = 60{,}000 + \left(\frac{5.00 - P}{0.50}\right) \times 11{,}000 \qquad \text{18-1}$$

For the average, the number of books, N , that can be sold versus unit price, P , is shown in Equation 18-2.

$$N = 60{,}000 + \left(\frac{5.00 - P}{0.50}\right) \times 10{,}000 \qquad \text{18-2}$$

For the lower 95% confidence limit, the number of books, N , that can be sold versus unit price, P , is shown in Equation 18-3.

$$N = 60{,}000 + \left(\frac{5.00 - P}{0.50}\right) \times 9{,}000 \qquad \text{18-3}$$

Since the total revenue, R, from sales of this book is $N \times P$, following parabolic equations, Equations 18-4, 18-5 and 18-6, are obtained for the upper 95% confidence limit, the average, and the lower 95% confidence limit, respectively.

$$R = -22{,}000 \times P^2 + 170{,}000 \times P \qquad \qquad \text{18-4}$$

$$R = -20{,}000 \times P^2 + 160{,}000 \times P \qquad \qquad \text{18-5}$$

$$R = -18{,}000 \times P^2 + 150{,}000 \times P \qquad \qquad \text{18-6}$$

A change in total revenue versus a change in book's unit price can be written in a finite difference form by taking derivatives of total revenue with respect to book's unit price in Equations 18-4, 18-5 and 18-6. Finite difference equations for the upper 95% confidence limit, the average, and the lower 95% confidence limit are shown, respectively, in Equations 18-7, 18-8 and 18-9.

$$\frac{R_{i+1}-R_i}{P_{i+1}-P_i} = -44{,}000 \times \left(\frac{P_{i+1}+P_i}{2}\right) + 170{,}000 \qquad 18\text{-}7$$

$$\frac{R_{i+1}-R_i}{P_{i+1}-P_i} = -40{,}000 \times \left(\frac{P_{i+1}+P_i}{2}\right) + 160{,}000 \qquad 18\text{-}8$$

$$\frac{R_{i+1}-R_i}{P_{i+1}-P_i} = -36{,}000 \times \left(\frac{P_{i+1}+P_i}{2}\right) + 150{,}000 \qquad 18\text{-}9$$

Optimum estimated total revenue versus book's unit price is calculated for three cases using exact parabolic Equations 18-4, 18-5 and 18-6 and finite difference Equations 18-7, 18-8 and 18-9. The results are presented in Figure18-1. In these cases, exact equations and finite difference equations provide same results, since the derivative, $\frac{dR}{dP}$, is a linear function of book's unit price. Therefore book's unit price could be represented by $\left(\frac{P_{i+1}+P_i}{2}\right)$ in finite difference equations.

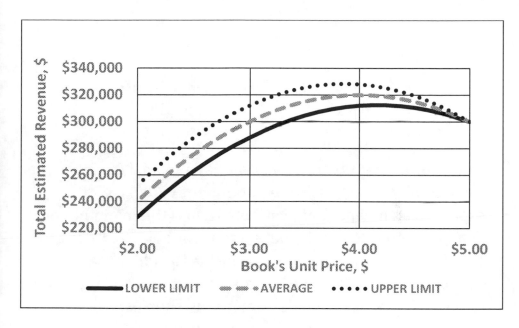

Figure 18-1: Optimum estimated revenue versus
book's unit price

In this marketing analysis, optimum estimated revenue point for the lower 95% confidence limit is $312,500 at the book's unit price at $4.17 and an estimated 75,000 books will be sold, for the average case, the optimum estimated revenue point is $320,000 at the book's unit price at $4.00 and an estimated 80,000 books will be sold, and for the upper 95% confidence limit, the optimum estimated revenue point is $328,409 at the book's unit price at $3.86 and an estimated 85,000 books will be sold.

Chapter 19

Shortest Ladder over a Wall

In most optimization problems, we have to analyze the first and second derivatives of the dependent variable with respect to an independent variable to find the minimum or the maximum value of the dependent variable. In order to illustrate such a problem, let us consider a ladder that has to lean on a wall in order to reach a structure on the other side of the wall. We will try to determine the shortest ladder length that is required to reach the structure as illustrated in Figure 19-1.

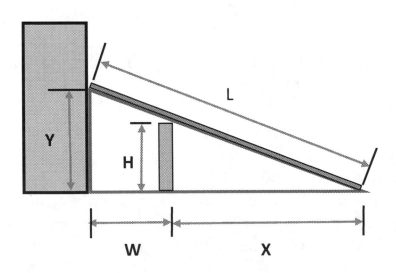

Figure 19-1: Ladder over a wall to reach the structure

H is the height of the wall. W is the horizontal distance from outside of the wall to the structure's face. L is the length of the ladder. X is the horizontal distance from outside of the wall at ground level to the ladder's back end where it touches the ground. Y is the vertical distance from ground level to where the ladder's front end touches the structure. All lengths are in meters. Length of the ladder can be defined using Pythagoras' theorem as shown in Equation 19-1.

$$L^2 = Y^2 + (W + X)^2 \qquad \text{19-1}$$

We can define Y in terms of X using similar triangles as given in Equation 19-2.

$$\frac{Y}{W+X} = \frac{H}{X} \qquad \text{19-2}$$

We can eliminate Y by combining Equations 19-1 and 19-2 and obtain the following Equation 19-3 for the dependent variable L as a function of only one independent variable, namely X .

This is the exact equation form for the ladder length, L, versus the independent variable X.

$$L = \left(\frac{1}{X}\right) \times (W + X) \times (H^2 + X^2)^{0.5} \qquad \text{19-3}$$

Since the wall height and the distance from outside of the wall to the structure's face are known, we can optimize the required ladder length with respect to the independent variable X by taking the derivative of L with respect to X in Equation 19-3. The result is shown in Equation 19-4.

$$\frac{dL}{dX} = \frac{X^3 - W \times H^2}{X^2 \times (H^2 + X^2)^{0.5}} \qquad \text{19-4}$$

From Equation 19-4, we can obtain the finite difference equation form for the ladder length versus the independent variable X as shown in Equation 19-5.

$$\frac{(L_{i+1}-L_i)}{(X_{i+1}-X_i)} = \frac{\left[\left(\frac{X_i+X_{i+1}}{2}\right)^3 - W\times H^2\right]}{\left(\frac{X_i+X_{i+1}}{2}\right)^2 \times \left[H^2+\left(\frac{X_i+X_{i+1}}{2}\right)^2\right]^{0.5}} \qquad 19\text{-}5$$

Now let us analyze a case to find a minimum ladder length where the height of the wall. H is 2 meters and the distance from outside of the wall to the structure's face W is 4 meters. A change in the length of the ladder versus a change in the X distance is calculated using both Equations 19-3 and 19-5. The results are shown in Figure 19-2. For approximate solutions, a small increment of $(X_{i+1} - X_i) = 0.01\ meter$ is used in order to achieve results that are in line with exact results. The ladder length goes through a minimum at $X = 2.515\ meters$ and the minimum ladder length comes out to be 8.32 meters. With the approximate solution, the minimum ladder length occurs at $X = 2.52\ meters$ and the minimum ladder length comes out to be 8.39 meters, namely with a 0.8% error.

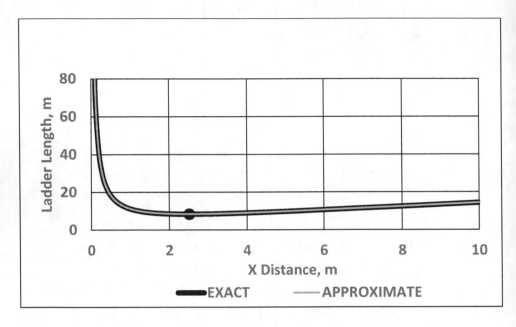

Figure 19-2: Ladder length versus X distance: exact and approximate solutions

At the minimum ladder length, the first derivative of L with respect to X has to go through zero. Using Equation 9-4, the first derivative of L with respect to X is plotted in Figure 19-3. $\frac{dL}{dX}$ goes through zero only once at an X value of 2.515 meters.

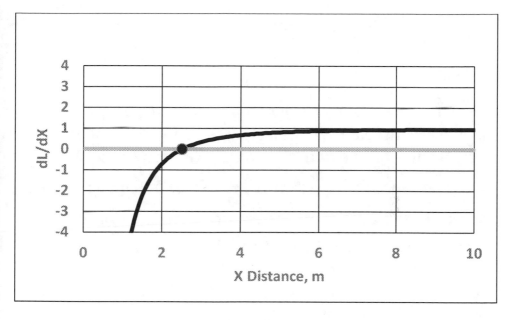

Figure 19-3: dL/dX versus X

The value of X that makes $\frac{dL}{dX} = 0$ can be obtained from Equation 9-4 in another way. We can observe that the denominator of Equation 9-4 cannot be zero since X cannot be zero. Therefore for $\frac{dL}{dX} = 0$ can be achieved only if the numerator is zero, namely $X_{\frac{dL}{dX}=0} = W^{1/3} \times H^{2/3}$. We can substitute this result into Equation 19-3 in order to get an expression for the minimum ladder length as functions of H and W as shown in Equation 19-6.

$$L_{minimum} = \left(W^{2/3} + H^{2/3}\right)^{1.5}$$

19-6.

We have to perform another step in order to verify that the ladder length given in Equation 19-6 is a minimum. For this to be true, the second derivative of L with respect to X has to be positive when $\frac{dL}{dX} = 0$. $\frac{d^2L}{dX^2}$ can be obtained by taking the derivative of $\frac{dL}{dX}$ with respect to X in Equation 19-4 which is presented in Equation 19-7.

$$\frac{d^2L}{dX^2} = \frac{3}{(H^2+X^2)^{0.5}} - \frac{(X^3-W\times H^2)\times(2\times H^2+3\times X^2)}{X^3\times(H^2+X^2)^{1.5}}$$

19-7

$\frac{d^2L}{dX^2}$ versus X is plotted in Figure 19-4. It is shown that the second derivative of L with respect X is always positive, including the point where $\frac{dL}{dX} = 0$.

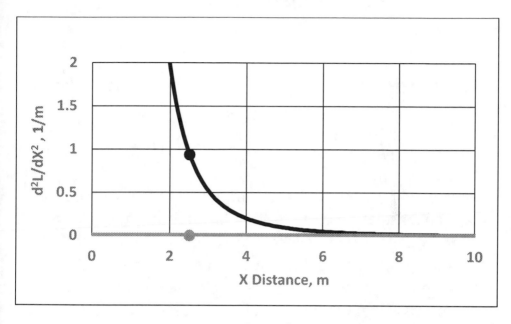

Figure 19-4: d^2L/dX^2 versus X

Chapter 20

Optimizing Heating and Insulation Costs for a House

While building a house, a combination of long term heating costs and insulation costs have to be analyzed in order to determine the optimum insulation thickness that will minimize long term heating plus insulation costs. As the insulation thickness in the walls and in the ceiling increase, heating loss from the house will decrease and therefore long term heating costs will decrease. At the same time thicker insulation means higher initial construction costs. In the present example, we will analyze how changing the thickness of insulation affects total costs for heating and insulating a house.

Let us assume that insulation costs increase linearly with its thickness. As a baseline, we will use 9 cm thick fiberglass insulation which is $10 per square meter. A house, with a 400 m^2 area for losing heat through its walls and ceiling, has to be insulated. Insulation costs versus insulation thickness is shown in Figure 20-1.

The insulation costs for this house increase linearly up to $22,222 at 0.5 insulation thickness.

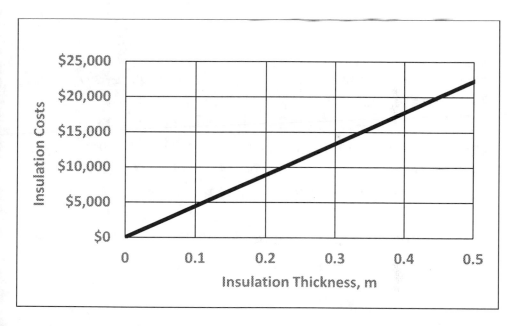

Figure 20-1: Insulation costs versus insulation thickness

Let us consider 80% efficient natural gas heating for this house for a 30-year duration. The house is in a mild climate environment and requires heating only 3 months out of a year, namely 64,800 hours of heating in 30 years. We will also assume that the natural gas prices will be at a constant rate during the 30-year period, namely at 0.1 $/kW-hr.

For the present value analysis, a constant rate gas price assumption allows the natural gas prices to increase as much as treasury bills interest rates which offset the annual inflation rates.

In the present analysis, we will consider one dimensional and steady-state heat transfer from air inside the house to air outside the house. Heat is lost by convection and by conduction through the insulated walls and the ceiling while neglecting any air leaks. Walls and the ceiling are made out of three layers of materials, namely the inner wall/ceiling board, the insulation layer and the outer wall/ceiling board. Heat is transferred per unit area from inside air to outside air is given by Equation 20-1.

$$\frac{Q}{A} = \frac{(T_{inside\ air} - T_{outside\ air})}{\left(\frac{1}{h_{in}}\right) + \left(\frac{t_{inner\ wall}}{k_{inner\ wall}}\right) + \left(\frac{t_{insulation}}{k_{insulation}}\right) + \left(\frac{t_{outer\ wall}}{k_{outer\ wall}}\right) + \left(\frac{1}{h_{out}}\right)}$$

20-1

where Q is the heat transferred in watts. A is the total surface area from which heat is transferred, namely $A = 400\ m^2$ in the present case. h_{in} is the natural convection heat transfer coefficient from the inside air to the inner surface of the inner board.

For the present case $h_{in} = 4 \frac{W}{m^2 - K}$. $t_{inner\ wall}$ is the thickness of the inner wall/ceiling board material at 0.02 m. $k_{inner\ wall}$ is the thermal conductivity of the inner wall/ceiling board material and has a value of 0.12 W/m-K. $t_{insulation}$ is the thickness of the insulation material. Insulation thickness is a variable in the present analysis that we will determine for minimum heating plus insulation costs. In the present analysis, the insulation thickness varies from zero to 0.5 meter. $k_{insulation}$ is the thermal conductivity of the insulation material and has a value of 0.046 W/m-K. $t_{outer\ wall}$ is the thickness of the outer wall/ceiling board material at 0.01 m. $k_{outer\ wall}$ is the thermal conductivity of the outer wall/ceiling board material and has a value of 0.72 W/m-K. . h_{out} is the combination natural and forced convection heat transfer coefficient from the outer surface of the outer board to the outer air. For the present case $h_{out} = 7 \frac{W}{m^2 - K}$. For the present analysis, $T_{inside\ air}$ is the average inside air temperature at 20 C and $T_{outside\ air}$ is the average outside air temperature at 5 C. Using Equation 20-1 and parameters given above, natural gas energy that is required to heat this house versus insulation thickness is shown in Figure 20-2. With no insulation, 10.74 kW-hr natural gas energy is needed per hour in order to keep the interior of the house at 20 C. The hourly natural gas energy requirement decreases inversely proportional to increasing insulation thickness, namely $\frac{Constant1}{(Constant2 + t_{insulation})}$ where $Constant1$ and $Constant2$ can be obtained from Equation 20-1.

When the insulation thickness is 0.5 meter, 0.6 kW-hr natural gas energy is needed per hour in order to keep the interior of the house at 20 C when the outside temperature is at 5 C.

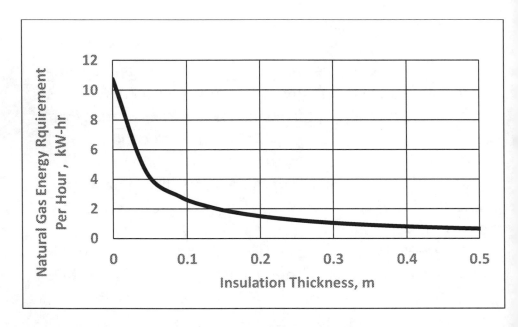

Figure 20-2: Required natural gas energy per hour
to heat the house versus insulation thickness

Using the results in Figure 20-2, with 64,800 required heating hours in 30 years and using the natural gas price of 0.1 $/kW-hr, 30-year heating costs can be determined. 30-year heating costs versus insulation thickness is plotted in   Figure 20-3. Behaviors of curves in Figures 20-2 and 20-3 are the same since we assumed constant natural gas prices over 30 years.

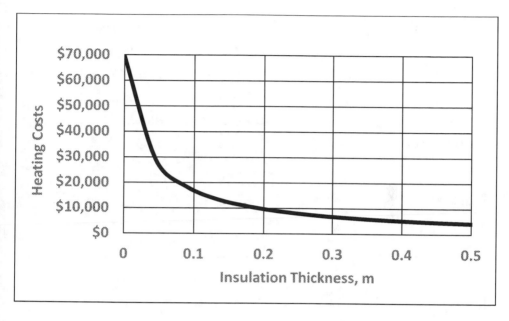

Figure 20-3: 30-year heating costs versus insulation thickness

Present value of 30-year heating costs versus insulation thickness and house insulation costs versus insulation thickness are presented above, respectively, in Figures 20-1 and 20-3. Figure 20-4 shows how heating plus insulation costs are varying as a function of insulation thickness, namely the sum of cost curves in Figures 20-1 and 20-3. With no insulation, heating costs are the highest, about $70,000, as expected. As the insulation thickness increases, heating costs decrease, but the insulation costs increase. As the insulation thickness increases, decreasing heating plus insulation costs go through a minimum and then they starts to increase again.

For the present case, the minimum heating plus insulation costs occur at an insulation thickness of 0.18 meter and it is $18,539. Increasing the insulation thickness above 0.18 meter, insulation costs will overtake heating costs and increase total heating and insulation costs as shown in Figure 20-4.

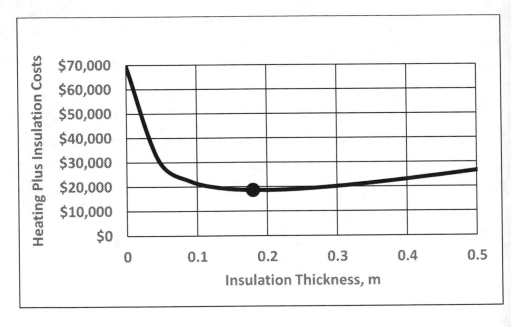

Figure 20-4: Heating plus insulation costs versus insulation thickness

Chapter 21

Expansion of a Vineyard for Maximum Profit

The quality of a wine starts in the vineyard. Vineyards are intentionally farmed to produce low yields so that they can get the best tasting grapes. In the present case, a vineyard owner has a 150 acre vineyard for growing grapes. From 150 acres of vineyard, on the average he can yield 2 barrels of regular wine per acre every year. So his wine production on the average is 300 barrels of regular wine per year. From his 150 acres of vineyard, in addition to regular wine, he can also yield on the average 0.5 barrel of premium wine per acre every year, namely on the average 75 barrels of premium wine per year.

He wants to add additional acreage to his vineyard in order to increase his good grape production and therefore his wine production in order to maximize his profits. However, historical data and studies for his region show that when a vineyard is expanded, usable good grapes for regular wine decrease by 0.5% per expanded acre. Also, historical data and studies for his region show that when a vineyard is expanded, usable excellent grapes for premium wine decrease by 0.3% per expanded acre.

From each barrel, he produces 300 bottles of wine, i.e. 750 ml bottles. He profits $1 per bottle from regular wine sales and $3 per bottle from premium wine sales.

With all the above inputs that he has, he wants to determine the vineyard acreage expansion that he should plan for in order to maximize his profits. This is a straightforward optimization problem to find out as how to enlarge his vineyard acreage while decreasing good and excellent grape yields in order to maximize his annual profits. His present acreage plus his planned acreage can be equated to barrels of wine production as shown in Equation 21-1.

$$NB = (PA + FA) \times [(2 - 0.005 \times 2 \times FA) + (0.5 - 0.003 \times 0.5 \times FA)]$$
21-1

where NB is the average number of total barrels of wine that can be produced every year. PA is the present vineyard acreage which is 150 acres. FA is the future vineyard acreage in acres. Equation 21-1 can be rewritten to determine number of barrels of regular wine and of premium wine that can be produced every year with additional acreage. See below Equation 21-2a and 21-2b.

$$NB_{regular} = (PA + FA) \times (2 - 0.005 \times 2 \times FA)$$

21-2a

$$NB_{premium} = (PA + FA) \times (0.5 - 0.003 \times 0.5 \times FA)$$

21-2b

His total profit from the vineyard acreage expansion is defined by Equation 21-3.

$$Total\ Profit = \$1 \times 300 \times NB_{regular} + \$3 \times 300 \times NB_{premium}$$

21-3

Wine production versus additional vineyard acreage is calculated using Equations 21-1 and 21-2. The results are presented in Figure 21-1. With additional acreage, total wine production initially increases, goes through a maximum at 80 additional acres and then starts to decrease due to dropping yields. Regular wine production initially increases with additional acreage, goes through a maximum at 120 additional acres and then starts to decrease due to dropping yields.

Similarly, premium wine production initially increases with additional acreage, goes through a maximum at 10 additional acres and then decreases to zero at 175 additional acres.

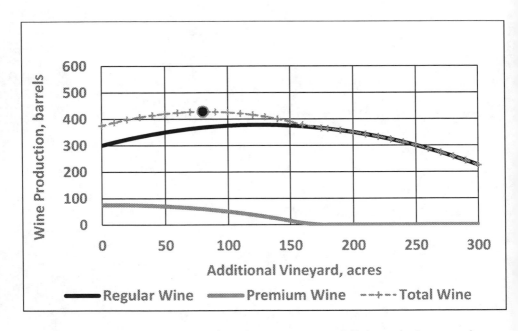

Figure 21-1: Wine production versus additional vineyard acreage

This vineyard owner's projected profits from the vineyard acreage expansion are calculated using Equation 21-3. Total profit per year versus additional vineyard acreage is shown in Figure 21-2. His present profit from 150 acres of vineyard is $157,500 per year in which he produces 300 barrels of regular wine and 75 barrels of premium wine.

His annual profit by adding additional vineyard acreage goes through a maximum at 50 additional acreage and it is projected to be $168,000 per year from production of 350 barrels of regular wine and 70 barrels of premium wine. His annual profits starts to decrease sharply after 50 additional acres.

The inflection point at 175 additional acres on the total profit per year versus additional vineyard acreage curve is due to the end of premium wine production.

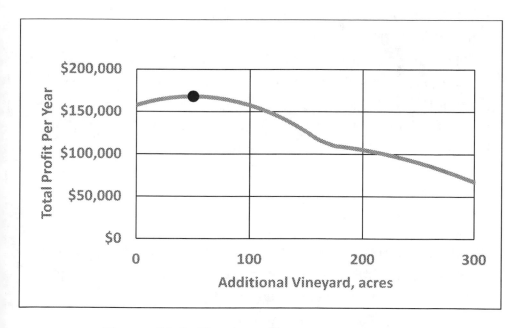

Figure 21-2: Total profit per year versus additional vineyard acreage

Chapter 22

Genetic Manipulation of Egg Laying Hens

Changes in a variable that is critical to a process can be identified by collecting a set of controlled data and then analyzing it statistically. In the present case, let us analyze an egg farm where eggs under a certain size are categorized as rejects and cannot be shipped to customers. The egg farmer records meticulously, on a daily basis, number of rejected small eggs and his daily total egg production. He wants to improve his production and his shippable egg yields by performing a genetic manipulation to his egg laying hens. He collects two months of data after the genetic manipulation to his hens. He is ready to analyze his pre and post genetic manipulation data for 60 days statistically to find out changes that happened in his egg production numbers, his small egg reject numbers and how genetic manipulation is going to affect his annual gross income.

Before the genetic manipulation, he produces 400,000 eggs per year from 2,000 hens, namely on the average 200 eggs per hen per year. He sells each his large eggs that are shippable to customers at $ 0.12.

The farmer uses the following Equations 22-1 and 22-2, respectively, to calculate the average of percentage of daily small egg rejects and the standard deviation of the percentage of daily small egg rejects.

$$\bar{X} = \frac{1}{60} \times \sum_{i=1}^{60} \left(\frac{R}{N}\right)_i \times 100 \qquad\qquad 22\text{-}1$$

$$\sigma = \sqrt{\frac{\sum_{i=1}^{60}(X-\bar{X})^2}{(60-1)}} \qquad\qquad 22\text{-}2$$

where R is the number of daily small egg rejects and N is the number of daily egg production.

His average daily small egg rejects before genetic manipulation comes out to be 17.1 %. The standard deviation for the 60-day data before the genetic manipulation is 1.94 %. On the average, he loses 17.1 % of his eggs and he is able to sell only 331,507 eggs per year with an average gross income of $39,781.

He also wants to verify that there are no extreme values and behavior in his 60-day data set that can distort values of averages and standard deviations. He sorts his daily small egg reject percentage data into equal class intervals and makes a histogram from the frequency of data in each class interval. Frequencies of daily small egg reject percentages for 60 days before genetic manipulation is shown as a histogram in Figure 22-1.The frequency distribution behaves close to a symmetric distribution, like a Gaussian distribution with an $\bar{X}$ of 17.1% and $\pm 3\sigma$ of $\pm 5.8\%$ and there are no extreme values that could affect his average and standard deviation calculations.

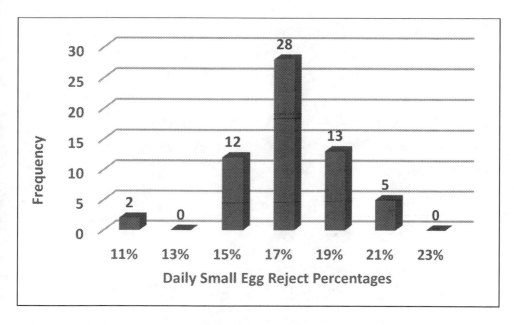

Figure 22-1: Frequency of daily small egg reject percentages before genetic manipulation

After the genetic manipulation, 60-day data collection shows that his egg production per hen per year increases to 250 eggs and losses due to rejected small eggs decreases. His average daily small egg rejects after genetic manipulation is 7.0 %.. The standard deviation for the 60-day data after genetic manipulation is 1.2 %. On the average, he loses only 7.0 % of his eggs and he is able to sell 464,967 eggs per year with increased annual per hen per year production rate. He is projected to have an average gross income of $55,796 after the genetic manipulation.

Frequency of daily small egg reject percentages for 60 days after genetic manipulation is shown as a histogram in Figure 22-2. The frequency distribution again behaves close to a symmetric distribution, like a Gaussian distribution with an $\bar{X}$ of 7.0% and $\pm 3\sigma$ of $\pm 3.6\%$ and again there are no extreme values that would affect his averages and standard deviations.

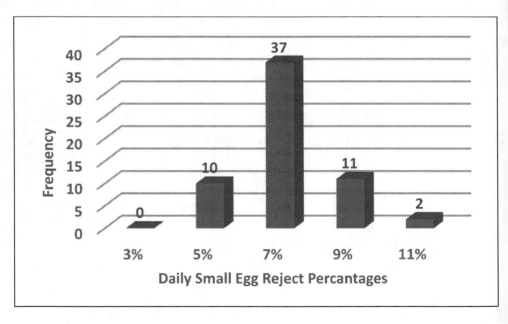

Figure 22-2: Frequency of daily small egg reject percentages after genetic manipulation

If Figures 22-1 and 22.2 are analyzed more carefully, we can see that the average number of daily small egg rejects is reduced by 59.1 % after genetic manipulation. Also the spread, i.e. standard deviation, of the daily small egg reject distributions improves by 40.1 % after genetic manipulation. Above all, his annual gross income increases by 40.3% after genetic manipulation.

Chapter 23

Changing Processes or Measurement Systems in High Volume Production

In high volume production, it is not feasible and cost effective to measure each part to see if it complies with a critical specification. First we have to have a capable measurement system for the critical specification in question in order to be able to measure samples in a production line. The measurement system's variability has to be less than 10% of our product's critical specification's limits range. Control charting critical specification by sampling can show you immediately, if your process is changing or it is out-of-control. Similar control charting applies to your measurement systems. You can see if your measurement systems are in control or if they have changed. You can make immediate decisions to shut a high volume production line down in order to find and correct the problem.

Variable control charts such as $\bar{X}$, i.e. average of the production samples, and R , range of the production samples, provide excellent information about the overall average and spread of a product with respect to a critical specification.

When your control charts for all your product's critical specifications are in statistical control and your critical specifications' averages and standard deviations all provide you with 3.4 defective parts out of a million part produced, i.e. $\pm 6\sigma$ process capability, then you can proudly say to your customers that you have a a product that meets all critical specifications.

For product variables, there are two type of popular control charts. One is sample average and sample range control chart. The other is (for large sample sizes, i.e. sample size > 10) sample average and sample standard deviation control chart. For product attributes, there are P , NP , C , and U control charts. See M. Kemal Atesmen's book on "Process Control Techniques for High-Volume Production", CRC Press, 2017, Reference [1]. In the present chapter, we will only investigate sample average, $\bar{X}$, and sample range, R , control charts.

$\bar{X}$ is the average of the critical specification for the product sample where the sample size is less than 10 and it is a constant. $\bar{X}$ for a product sample size of 5 is given by Equation 23-1.

$$\bar{X} = \frac{X_1 + X_2 + X_3 + X_4 + X_5}{5}$$ 23-1

R is the range of the critical specification for the product sample as shown in Equation 23-2.

$$R = X_{\max \, for \, sample \, group} - X_{\min \, for \, sample \, group}$$ 23-2

In our example, the plating layer thickness for insulation is critical on a chip wafer. Two hundred wafers go through the plating process every hour. The operator pulls out five wafers at random every hour and measures the critical plating thickness. $\bar{X}$ and R control charts are shown for this critical parameter for 24 hours on 24 September 2014 in Figures 23-1 and 23-2, respectively.

Process average of all samples have been calculated from the initial 30 sample lots' averages at the beginning of this chip's wafer production and it is 24.5 µm. Similarly process average of ranges have been calculated from 30 sample lots' ranges and it is 3.7 µm. Control limits for the Charts in Figure 23-1 and 23-2 are calculated by using a sample size of 5. See M. Kemal Atesmen's book on "Process Control Techniques for High-Volume Production", CRC Press, 2017, Reference [1], for control chart limit calculations. $\bar{X}$ control chart's upper control limit comes out to be 26.6 µm and lower control limit comes out to be 22.3 µm. . R control chart's upper control limit comes out to be 8.5 µm and lower control limit comes out to be 0.0 µm. The plating thickness process seems to be in statistical control. However, we have to perform a process capability analysis in order to be able to assure our customers that our production process is capable to $\pm 6\sigma$.

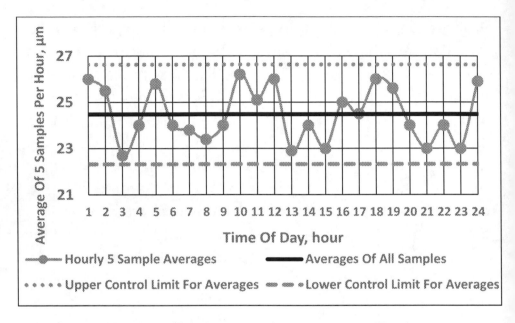

Figure 23-1: Averages control chart for plating thickness versus production hours on 9/24/2014

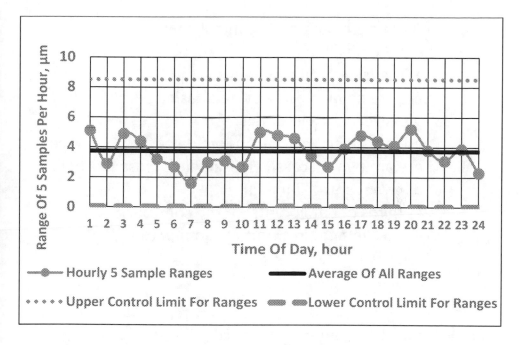

Figure 23-2: Ranges control chart for plating thickness versus production hours on 9/24/2014

Both sample averages and ranges control charts shown above are in statistical control because we do not come across any of the following process out-of-control conditions in both control charts.

1. No points are outside of control limits.
2. There are no eight points of data runs in a succession on either side of the averages.
3. There are no increasing or decreasing data trends successively in six intervals.
4. There are no cyclic non-random patterns.
5. There are two successive data points close to upper and lower control limits.
6. Two-thirds of data points on the control chart are not close to averages in a non-random pattern, i.e. no variability in data.

If any of the above out-of-control conditions appear on a control chart, the operator has to act immediately to stop the process and to correct for the changing process or for the changing measurement system.

For the plating thickness example treated above, we can estimate the process standard deviation from the mean of sample ranges as shown in Equation 23-3.

$$\sigma = \frac{\bar{R}}{d_2} = 1.6 \ \mu m \qquad\qquad 23\text{-}3$$

d_2 depends on sample size and it is 2.326 for the present case's sample size of 5. See M. Kemal Atesmen's book on "Process Control Techniques for High-Volume Production", CRC Press, 2017, Reference [1], for details. We can show to our customers that our plating thickness process is capable to $\pm 6\sigma$, namely

$$\bar{X} + 3 \times \sigma = 34.1 \ \mu m \qquad \text{and}$$

$$\bar{X} - 3 \times \sigma = 14.8 \ \mu m$$

for the critical plating thickness specification of $24.5 \pm 10 \ \mu m$.

Chapter 24

Probability of Warranty Return Defects

An event can be characterized by the probability of its happening. Changes in the probability of an event can provide crucial information for decision making. If there are two or three or more number of events that are not related, namely mutually exclusive, probability of each event occurring can be added together to get the probability of all events occurring that are under consideration.

For example, if an appliance store sells brand X washing machines from vendor A and brand Y refrigerators from vendor B. After a year of sales, the appliance store manager analyzes all warranty return data. Brand X washing machines has a 6% warranty returns to total sales in a year. Brand Y refrigerators has a 4.5% warrant returns to total sales in a year. If both vendors do not improve on any of the warranty return defects, what will be the projected total probability of warranty returns from these two products in the coming year, namely total warranty returns from brand X washing machines or brand Y refrigerators?

Since brand X washing machine warranty returns and brand Y refrigerator warranty returns are mutually exclusive occurrences, total probability of warranty returns from these two products is shown in Equation 24-1.

$$P(X \text{ or } Y) = P(X) + P(Y) = 6\% + 4.5\% = 10.5\%$$

24-1

In another example, a large electronics store sells personal computers from vendor C. After a large number of personal computer sales from this vendor C, the store has the following percentages for warranty returns to total sales data:

Warranty returns due to disk drive, (DD), failures = 0.16%

Warranty returns due to keyboard, (KB), failures = 0.07%

Warranty returns due to disk drive, (DD), and keyboard, (KB), failures = 0.05%

These two warranty return events are mutually non-exclusive. Therefore probability of first event is added to the probability of the second event and then probability of two events occurring together is subtracted. The probability of warranty returns for personal computers from vendor C is given in Equations 24-2a and 24-2b.

$$P(DD \ or \ KB) = P(DD) + P(KB) - P(DD \ and \ KB)$$

24-2a

$$P(DD \ or \ KB) = 0.16\% + 0.07\% - 0.05\% = 0.18\%$$

24-2b

In another example, this large electronics store also sells personal computers from vendor D which is a preferred vendor with ship-to-stock status. After a large number of personal computer sales from this vendor D, the store has the following percentages for warranty returns to total sales data:

Warranty returns due to disk drive, (DD), failures = 0.04%

Warranty returns due to keyboard, (KB), failures = 0.06%

Warranty returns due to disk drive, (DD), and keyboard, (KB), failures = 0.03%

These two warranty return events are again mutually non-exclusive events. The probability of warranty returns for personal computers from vendor D is given in Equation 24-3.

$$P(DD \text{ or } KB) = 0.04\% + 0.06\% - 0.03\% = 0.07\%$$

24-3

Probability of warranty returns from vendor D is 61% less than probability of warrant returns from vendor C. The electronics store has to get after the personal computer vendor C in order for them to get their act together.

References

[1]   Atesmen, M. K., "Process Control Techniques for High-Volume Production", CRC Press, New York, 2017.

[2]   Boyce, W. E. and R. C. DiPrima, "Elementary Differential Equations and Boundary Value Problems", John Wile & Sons, Inc., New York, 1997.

[3[Halliday, D., R. Resnick and K. S. Krane, "Physics", John Wile & Sons, Inc., New York, 1992.

[4]   Kreith, F., ""Principles of Heat Transfer", International Textbook Company, Scranton, Pennsylvania, 1965.

[5]   Levin, R. I., "Statistics for Management", Prentice-Hall, New Jersey, 1987.

[6]   Shames, R. H., "Mechanic of Fluids", McGraw-Hill, New York, 1962.

Index

Decay, V, VII, XV, 15, 88-91,
Density, XV, 9 67, 69, 78-79,
     83-84, 93
Depth, V, X, XV, 83-86
Dependent variable, V-VI,
     X, XVI-XVII, 108-110
Difference, V, X, 1, 5, 8, 11-13
     15, 17 , 49-58, 61-64
Differential equation, V, X, XI,
     5, 12, 17, 57-58, 63,
     89, 146
Digital, VI
Discharge, II, 49, 52-54, 92
Disease, II, XIII, 55-60
Distance, XIII, 23-26, 36-42,
     109-112, 113, 115,
Distance, XIII, 23-26, 36-42,
     109-113, 115

E

Education, II, 1
Environment, IX, XII, 8-9,
     79, 117
Epidemics, XIV, 72
Evaporation, III, XIV, 61-64
Experiment, X, 9, 15, 92

F

First order, V, X, XV, 1,
     5, 8, 12-13, 15, 17,
     50, 52-53, 57-58,
     63, 88-89
Forced convection, 119
Fund, XI, 1-7

G

Genetic, III, XVIII, 128-133
Genome, V
Gravitational acceleration, II,
     21, 30, 32, 38, 43-44,
     79, 84, 93
Growth, III, 72, 74-75, 77,
     99-102

H

Heat flow, XII, 8
Heat transfer, IX, XI, XIX, 9,
     11, 118-119, 146
Historical data, VI, X, 123
Hole, III, XVI, 92-93, 96, 98
House, III, XVII, 116-121

Printed in the United States
By Bookmasters